Seeing the City Digitally

Seeing the City Digitally

Processing Urban Space and Time

Edited by
Gillian Rose

Routledge
Taylor & Francis Group
LONDON AND NEW YORK

The publication of this book is made possible by a grant from the University of Oxford.

First published in 2022 by Amsterdam University Press Ltd.

Published 2025 by Routledge
4 Park Square, Milton Park, Abingdon, Oxon OX14 4RN
605 Third Avenue, New York, NY 10158

Routledge is an imprint of the Taylor & Francis Group, an informa business

ISBN: 9789463727037 (hbk)
ISBN: 9781041185895 (pbk)
ISBN: 9781003703464 (ebk)
NUR 670

Cover illustration: iStock

Cover design: Coördesign, Leiden

DOI: 10.5117/9789463727037

Table of Contents

Acknowledgements

Many of the chapters were presented as papers at a series of events hosted at St John's College, Oxford in 2018 and 2019. I would like to thank St John's College and the School of Geography and the Environment, University of Oxford, for their generous funding, and all the speakers and audiences who participated. Adam Packer, Alice Watson and Oliver Zanetti helped to programme one of the events, and I'd particularly like to thank Sterling Mackinnon III for all the work he did to organise and publicise the series.

Chapter 4, Ayona Datta

A longer version of this chapter was published in *Cultural Geographies*, titled "Curating #AanaJaana [#ComingGoing]: Gendered Authorship in the 'Contact Zone' of Delhi's Digital and Urban Margins" (Datta and Thomas 2021). This publication has been made possible through an AHRC funded research network (PI ref: AH/R003866/1). It would not have been possible without the participation and enthusiasm of the young women in the low-income settlements in Delhi who were part of this research. I am also grateful to Arya Thomas, our local research assistant for her hard work and support in interviews and organising community workshops. I am also grateful to Jagori and Safetipin for partnering on this project, and providing support and access to the community, recruiting participants, organising the workshops and feedback on this research. The exhibition panels were designed by graphic artist Kruttika Susarla.

Chapter 5, Monica Degen and Isobel Ward

We would like to thank our interviewees who made this research possible. The article was funded by the Brunel University Research Development Fund.

Chapter 6, Gillian Rose

A shorter version of this chapter appears in the book *Spatial Transformations: The Effect of Mediatization, Mobility, and Social Dislocation on the Re-Figuration of Spaces*, edited by Angela Million, Christian Haid, Ignacio Castillo Ulloa and Nina Baur and published by Routledge in 2021.

Chapter 8, Zlatan Krajina

The author thanks Zagreb's City Office for Economy, Energetics and Environment protection for their assistance in gathering data about the 2020 earthquake.

Chapter 9, Scott Rodgers

The chapter is based on the research project "Planning, Participation and Social Media Platforms" which was funded by the EPSRC. The Principal Investigator was Susan Moore at University College London, and alongside myself the other co-investigator was Andrea Ballatore at Birkbeck, University of London, who I must also thank for helping prepare the image data sets used in this chapter.

1. Introduction: Seeing The City Digitally

Gillian Rose

Abstract
The argument that many cities are now digitally mediated is an increasingly familiar one. The social, experiential and physical spaces of a city are more and more often designed, defined, navigated and experienced with digital data shared with platforms. But from its app icon to its interface to its advertising campaigns, every platform deploys a wide range of imagery, and most successful social media platforms are based on sharing images. This book explores what's happening to ways of seeing urban spaces in the contemporary moment, when so many of the technologies through which cities are visualized are digital. The introduction explores how the processuality of digital images, and their near-ubiquitous circulation, are reconfiguring the spatial and temporal organization of urban life.

Keywords: mediation, platform, processuality, representation, animation

Introduction

This book explores what's happening to ways of seeing urban spaces in the contemporary moment, when so many of the technologies through which cities are visualized are digital. It is by no means comprehensive. Its chapters all explore specific examples of different kinds of digital technologies and examine different sorts of images in different cities: many other technologies, images and cities could have been their focus. However, cumulatively the chapters suggest some of the most important ways in which seeing urban spaces through digital devices is reconfiguring both how cities appear and what happens there.

The argument that many cities – perhaps all cities, in different ways – are now digitally mediated is an increasingly familiar one (early statements include Boyer 1996; Manovich 2006; Mitchell 2003). McQuire (2016, 1), for

Rose, G. (ed.), *Seeing the City Digitally. Processing Urban Space and Time.*
Taylor & Francis Group, 2022
DOI 10.5117/9789463727037_CH01

example, concludes his discussion by identifying the extension of digital networked media throughout urban space as "one of the key features distinguishing twenty-first-century urban experience from earlier modes of urban inhabitation". The social, experiential, and physical spaces of a city are more and more often defined, navigated, and experienced with data generated by digital devices. Software-enabled technologies work with digital data of many kinds, in a huge array of urban infrastructures and institutions. Data is generated, integrated, and analysed by various human and algorithmic agents, with consequences for things as diverse as the allocation of housing and healthcare, traffic management, policing, and the provision of infrastructure and services (see for example Anthopoulos 2017; Aurigi and Willis 2020; Eubanks 2017; Graham 2005; Marvin, Luque-Ayala, and McFarlane 2016; Willis and Aurigi 2018). Smartphones and their cameras and apps mediate more and more of everyday urban life, from socializing to travelling to eating. For Kitchin and Dodge, this means that cities must be understood in part at least through the organizational geometries of "code/space": "code/space occurs when software and the spatiality of everyday life become mutually constituted" (2011, 16).

Much recent discussion of code/space in urban studies has centred on the generation and integration of digital data for urban planning and city management. This was the focus of early accounts of "informational" and "intelligent" cities (see for example Batty 1990; Castells 1989) and it has remained central to much of the recent extensive discussion of the "smart city". In these discussions, a lot of attention has been given to how city authorities install and utilize digital infrastructure and data flows. The close relationship between digital infrastructure and the neoliberal privatization of city governance was noted early on and continues to be the focus of much criticism (Cardullo and Kitchin 2019; Hollands 2008). More recently, understanding the digital mediation of cities has had to engage with corporate digital platforms like AirBnB, Facebook, Instagram, and a plethora of ride-sharing and food-delivery apps, among many others. These platforms also do what smart cities purport to do: gather data, integrate data, and put data to use. However, while much smart city activity retains at least some relation to the forms and ideals of civic governance – even if only lip service – platform urbanism is largely driven by the search for profit (Cowley, Joss, and Dayot 2018; Sadowski 2020). Platforms are owned by companies making money from vast, globally-integrated data assets and their machine-learning algorithms (Barns 2020a; Hodson et al. 2020).

Most analyses of these infrastructures and platforms have focussed on their extraction and commodification of the data generated by the platforms' users. However, some discussions have also begun to consider how platforms are also shifting the experiencing of urban life, as urban dwellers shop, play, eat, communicate, and work through them. "These platforms are, increasingly, the platforms many urban lives are increasingly constituted *by*" (Barns 2020a, 13). After all, a critical element in a platform's data infrastructure, and in many smart city projects, is a smartphone application, and apps are "functional and sensorial prostheses" for very many bodies (Srnicek 2014, 83). At the smartphone interface, platforms exert their pull, attuning users to their real-time, local connectedness; they are designed to be affectively trustworthy, seductive and effortless (Ash et al. 2017; Leszczynski 2019). Barns for example discusses the "intimate entanglements" between platforms and everyday urban life (2020a, 157).

Many of those entanglements are experienced visually. From its app icon to its interface to its advertising campaigns, every platform deploys a wide range of imagery. Indeed, the most successful social media platforms are based on sharing images: Whatsapp, Instagram, TikTok, Snapchat, even Facebook, and of course Pinterest. Intelligent and smart cities too rely on many kinds of visualizations, from the screens of smart city operations centres to online data dashboards to publicity displaying the benefits of going smart to their own smartphone applications (Luque-Ayala and Marvin 2016; Rose 2018; Rose et al. 2020).

And there are many other kinds of digital images in cities that picture urban spaces. Having been designed onscreen, new buildings are visualized in photorealistic detail in computer generated images, which appear on billboards and magazine advertisements, as well as on websites. Movies and computer games show cities that burn and fold and glow and float, cities that trundle along on huge caterpillar tracks, and cities that are flooded or frozen and sometimes both – or in ruins and inhabited by zombies or aliens. More prosaically, streets and cities are navigated using Google Maps (Wilmott 2016), and augmented reality apps – from games to local history projects – overlay smartphone cameras' view of roads and parks with other imagery (Uricchio 2011). Genres and purposes mix and blur as images circulate through any number of forms and places of display. Images of urban spaces are pasted or printed onto billboards and hoardings, flyers and brochures, magazines and newspapers; and urban spaces have been filling with screens large and small for some time (Manovich 2006). These extensive and diverse forms of digital imagery have been given relatively little attention in the work on digitally mediated cities. But they are central to how cities are changing now, and to how contemporary urban life is imagined. This book explores how digital images constitute urban code/spaces.

How do images of cities matter?

The argument that images of cities shape how cities are experienced is well established (Lindner and Meissner 2019). "The city is both the actual physical environment and the space we experience in novels, films, poetry, architectural design, political government, and ideology", notes Prakash (2008, 7), and to that list we could add photographs and maps and many other kinds of images. There is a productive "traffic between" cultural texts, everyday experiences and the urban built environment (Donald 1999, 27), so that the city becomes "the cognitive and somatic image we carry within us of the places where we live, work and play" (Huyssen 2008, 3). These arguments often emphasize the visual impact and discursive meaning of images. Images, it is argued, provide particular symbolic and affective co-ordinates for the experiencing of urban space. Images thus have their own liveliness.

But this argument must be pushed further. The mediation of urban life by images is not shaped simply by the visual content of the image and its impact on the imagination of its spectator. Images are never just visual content, whether symbolic or affective. Images take form as objects, and as objects they have material qualities (Rose and Tolia-Kelly 2012). As all of the chapters in this book point out, those qualities are variously mobilized, or not, by the socio-technical relations enacted as images are produced, reproduced, displayed, transported, modified, stored, and destroyed. Different kinds of images are made using different technologies in different ways; they are assembled and interpreted with other objects; and are seen, shared, and done other things with in various ways, with various effects (Clark 2018; Packer and Wiley 2012; Parks and Starosielski 2015; Pinney and Peterson 2003; Rose 2010). These makings and doings are routinized as social practices. As well as their visual content, these material affordances and practices are also part of an image.

The relation between an image – or imagery more generally – and urban space thus has a number of different elements. There are the visualizing technologies and the material affordances of image-objects, including what they picture of cities and how. There are the technologies of their distribution and the situations of their display. There are the social relations and institutions in which all of these are embedded, including how they are seen. There are the cultural meanings and significance on which they draw, or resist, and there are the affects that linger through all of these. There are thus co-constitutive relations between cities as sites of symbolic and affective images, and cities as sites of social practices and technologies.

There is a reciprocity between the material, social and symbolic forms of urban visualization and the visual perception of a city (Gordon 2010).

What then makes *digital* visualizations of urban spaces distinctive? After all, many digitally-produced images of cities look very similar to analogue images. Very many digital photographs look more or less the same as an analogue snap; for example, part of Instagram's early appeal was its ability to make a digital photograph look like a Polaroid. Much of the post-production work of big budget films now entails inserting the visual elements of analogue film into the digital movie (Murphy and Walker 2019). However, if we think of a digital visualization not simply as an image but as a lively socio-technical object embedded in socio-technical networks, as just described, then particular qualities of digital visualizations become evident, and their somewhat specific forms of configuring cities become more obvious. As the chapters in this collection propose, digital visualizations are doing something distinctive in their mediation of city space.

Visual technologies, practices, spaces

According to Besse (2013), by the end of the nineteenth century, in Europe for sure, visual culture was thoroughly urban. Cities were the sites of all sorts of innovations in visual technologies: balloon flights, panoramas, electric billboards, films, dioramas, photography, gas lighting. City dwellers bought new visual objects like daguerrotypes, postcards, tourist guides, and cartes de visite, and encountered new visual experiences in cinemas, arcades, expositions, and department stores.

As cities grew through the nineteenth and early twentieth centuries, various visualization techniques were also central to how cities were planned and managed. By the 1880s, there was a widespread assumption that the city must be made visible in order to be understood and managed. It had to be legible and "inspectable" (Otter 2008, 109). Otter (2008) traces the multiple, diverse, often hesitant, and indeed ineffective forms taken by practices of inspectability in European cities, from gas lighting to labelling to portable measurement devices. Many of these practices entailed creating images. Planners and social campaigners mapped, filmed, photographed and diagrammed both what needed improving as well as their ideal models for houses, neighbourhoods and cities. Much of this visualizing work was an effort to produce accurate evidence on which urban reform could be predicated (Barns 2020b; Boyer 1994; Clark 2018), and asserted an "absolute correspondence between the exterior city reality and its truthful and purified representation" (Boyer 1994, 19–21). Tagg (1988) explores the enrolment of photography into this project, as a mode of generating apparently reliable evidence about the need to improve urban housing, for example, and photographic technologies were used extensively in projects of urban planning and reform: with their technological indexicality and the assumption that they pictured the world objectively, "photographs seem to bond image to referent with superglue" (Mitchell 1992, 28). Entertainment technologies also trained their gaze at cities, as early filmmakers set up cameras in streets or on rooftops. The city – actual, desired and feared – thus became thoroughly visualized, and through these various forms of imagery, the urban environment was produced in particular ways. As Barns concludes:

> Through its entanglements with the evolution of urban planning, we can see the role of urban media as not simply representing the diverse conditions of urban transformation, but as helping to constitute the very production of urban space. This history sheds a different light on the nature of urban media technologies, suggesting it is not so much that urban media – whether those of historical eras or the smart technologies of more recent times – finally capture the true complexity of cities, but rather that they recalibrate urban knowledge and expertise in their own image. (2020b, 236–37)

Barns emphasizes in particular the importance of representational correspondence between the image and the real, in much of the imagery that was part of nineteenth and twentieth century urban management and planning. Projects to modernize cities took images of cities as imprints or traces of actual urban spaces.

Many of these sociotechnical practices of visually representing cities were challenged from the 1970s onwards. In part this was because planning itself was increasingly criticized as the best tool for managing urban life. However, the 1970s also saw the first sustained efforts at developing intelligent cities, based on a cybernetic understanding of a city not as one thing to be mapped, photographed and managed as a whole, but rather as a set of system of interconnected systems (Halpern 2015). This was the predecessor of the smart city. In it, subsystems, networks, and flow seemed to replace the visual insight of both the cartographic overview and more local practices of inspection. For some commentators, this made the intelligent city hard to see. Although total inspectability was a goal never successfully achieved (Flint 2000; Otter 2008), the intelligent city seemed especially difficult to visualize. An early commentary by Boyer on what she called "cybercities" suggested that this was because of what she considered the immateriality, incoherence, and extensivity of digital networks: "the whole has gone to pieces and no longer has imageable form" (1996, 175). Like Boyer, Barns (2020b) too suggests that cities have now reached some kind of limit of visibility, though she attributes this to the opacity of platforms' data harvesting and processing procedures to their users (Barns 2020a).

What the chapters in this collection contend, however, is that imagery still matters to urban code/space – but it is now often a different kind of imagery, with different consequences for the mediation of urban knowledge and experience. The materiality of this imagery is no longer analogue but digital, and its institutional context is less civic urban planning and management – though that remains important in many versions of smart cities – and more the "smart" platform urbanisms briefly described at the start of this introduction. In other words, dominant forms of urban imagery are now produced in the context of the material and corporate infrastructures of platform urbanism. To understand the implications of this shift, I suggest we need to turn away from urban planning as the context for understanding how cities are seen digitally, and towards recent discussions of digital cinema, digital photography, and even digital data visualization.

Discussions of digital cinema, digital photography, and digital data visualization clarify the distinctiveness of digital imagery by comparing it to analogue film, photography and data visualization (while being careful to avoid positing wholesale change driven by technological innovation) (see for example Casetti 2015; Elsaesser 2013a; 2013b; Denson and Leyda 2016; Halpern 2015; Levitt 2018). This comparison helps to specify how, although digital data in smart and platform cities continues to claim a certain verisimilitude to city spaces and urban life (Halpern 2015), there are nonetheless significant differences between the visual regimes through which nineteenth and twentieth century cities were seen and how intelligent, smart, and platform cities are visualized now. At the considerable risk of over-generalizing, for much of the nineteenth and twentieth centuries, urban imagery such as film, photography, maps, and diagrams was seen and deployed as representational. Representationalism is the conviction that what is represented exists independent of all practices of representation (Kember and Zylinska 2012, 31). Representational ways of seeing assume that there is a real that images – no matter how selective and distorted – re-present to the viewer. In film scholarship, this representationalist visual culture centred on lens-based recording has been called "cinematic". Those discussions focus on movies but could also refer to much urban photography as well as the visual tools of the planning profession as discussed by Barns (2020b). As for the effects of representational cinematic films:

> [O]ne would first list the *impression of reality*, that is to say, the high iconic fidelity that the photographic image carries. The 'reality-effect' is also a consequence of the *impression of movement*, which, in turn, is complemented by the *impression of presence*, strengthened by sound, but also providing one of the typical subject effects of cinema; namely, the impression of being included in the image and endowed with a special kind of *ocular-sensory, embodied identity.* (Elsaesser 2013a, 32)

What we see when we see cities cinematically, therefore, are representational images taken to refer to an external reality. It is true that describing this as cinematic conflates "the cinema as an audiovisual storage medium for motion pictures with the cinema as a projection-based spectacle in a public space" (Elsaesser 2013a, 26) – but given the importance of an image's materialization in socio-technical settings, this is a conflation that makes sense.

Thus defined, the cinematic is the dominant visual regime of modern city planning. It is a way of seeing "what happens" in urban spaces, and Asli Duru's chapter here is written against that grain. Duru's discussion also emphasizes that representationalist ways of seeing cities do not only constitute what the city is seen to be, but they also constitute particular kinds of observers. "Though obviously one who sees, an observer is more importantly one who sees with a prescribed set of possibilities, one who is embedded in a system of conventions and limitations" (Crary 1990, 6). As noted, Elsaesser describes the observer of film as endowed with "a special kind of *ocular-sensory, embodied identity*" (2013a, 32). The observer constituted by the films, photographs and maps and diagrams of planners tends to analyse. Their professional vision focusses on what needs managing and improving: it thus is always a powerful gaze that differentiates between good and bad kinds of urban spaces. As Otter (2008) discusses, inspectability was a visual regime trained much more comprehensively (though never anywhere near completely) on the poorest parts of nineteenth-century urban agglomerations. Particular bodies came under more scrutiny than others too: women's bodies and black bodies especially (for example, Browne discusses the eighteenth-century "lantern laws" in US cities which forced black, mixed-race and indigenous people to carry lights after dark, thus marking them as "security risks in need of supervision" [2015, 78]).

In contrast, many images in the twenty-first century neither depend on representationalism nor assume an analytic or supervisory eye. These images and their viewers have been described as "post-cinematic" by film scholars. Elsaesser makes this comparison, again in the context of movies:

> The key *digital effects* [are] the *impression of hyper-reality*, which would lead to an *impression* not of movement but *of metamorphosis*; that is, not only in the form of morphing and shape-shifting, but also as a constitutive instability of scale, mobility of point of view, and inherent 'liquidity' of the (visual) representation. Second, instead of giving an impression of identity and presence, provided in the cinema by the stable configuration of projection, frame, and linear fictional narrative, the subject effect typical of the digital would be the *impression of agency, tactility, and interactivity.* [...] In each case and on both sides of the divide, these effects are 'illusory'. (2013a, 33)

Various accounts of post-cinema concur (see for example the position statements gathered by Denson and Leyda [2016]). Mitchell (1992) describes post-photography in similar terms: post-photographs no longer imply presence. And while these analyses focus primarily on the visual or aesthetic effects of digital images, it is important also to acknowledge that these are entangled with – though not reducible to – distinctively digital forms of distribution and viewing as well as (post)production processing. Post-cinematic films and post-photographic photos – as well as all sorts of other images – are viewed on all sorts of screens and in all sorts of situations (Casetti 2015), many of them circulating from platform to platform as they are distributed by their makers, users, fans, modders, doomscrollers, producers, likers, and retweeters among others.

This broad-brush account of cinematic and post-cinematic ways of seeing cities obviously glosses over any number of nuances and complexities (many are explored in relation to film and cinema by Elsaesser [2013a]). The reality effect of photographs could be put to work to challenge the aerial viewpoint of the planner, for example, as strong traditions of documentary and community photography attest (see for example Stacey 2020). As for moving urban images, McQuire reminds us that "for Benjamin, film assumed epochal significance insofar as its characteristic organizing logic – based on fragmentation and reassemblage of appearances through montage – might enable citizen-viewers to grasp patterns of urban life that otherwise resisted embodied experience" (2020, 17). Post-cinematic effects can be achieved using analogue technologies, and analogue technologies mimicked by digital. Nonetheless, as Boyer (1996) and Barns (2020a) indicate, the visual mediation of digitally-saturated cities does not seem to have developed from the visualizations of nineteenth and twentieth century cities by planners and architects: "contemporary developments are contributing to the undermining of the representational paradigm" (McQuire 2016, 5), even if partially in all sorts of ways. So the comparison between cinema and post-cinema (or photography and post-photography) is a useful one, if only heuristically. Accounts of post-cinema and post-photography allow us to think about how digital images of cities look different, feel different, and organize urban space and time differently. The next section explores how the chapters in this book specify that difference.

Digital visual processing of urban space and time

So how might discussions of post-cinema help to understand how contemporary cities are being visualized in distinctively digital ways? The chapters in this book all explore the implications of digital images of cities' two key socio-technical affordances: they are *processed* data, and that processing creates images that *circulate*. Both those affordances create particular ways of seeing urban space and time.

Digital images are assembled from various combinations of data, software and hardware. The (nearly) infinite adaptability of digital images – the modifiability of their data and their ability to materialize in different forms – is a quite different visual affordance from analogue images. Elsaesser (2013a, 36–37) describes how digital images emerge from data that is harvested from the world and then manipulated by combinations of hardware and software, and suggests that this shifts digital images away from cinematic notions of representational capture, in which the image is seen as a trace or an imprint of the world, into something more akin to an ongoing process of extraction from and sculpting of the world. Rather than a representational *trace*, digital images are more like ongoing *events* (McQuire 2016, 5). Thus "post-cinematic images are thoroughly *processual* in nature, from their digital inception and delivery to their real-time processing in computational playback apparatuses" (Denson 2016, 194). Mitchell (1992, 7) similarly emphasizes the processing of data in his account of post-photography, and Halpern's (2015) account of the constitution of "beautiful data" in post-war cybernetics also describes the importance of dynamic interactions with data.

This account of post-cinema in particular has encouraged some scholars to point to animation as a visualizing technique which has historically used analogue tools but is in many ways a precursor of post-cinematic effects (and whose long history has been ignored in many classic accounts of cinema that define the cinematic as representational [Ristola 2017]). Animation moves imagery from "questions about ontology, category, and being to ones of appearance, metamorphosis, and affect" (Levitt 2018, 2). While not directly determined by technological changes, animated imagery is particularly enabled by the software that processes digital images, especially moving images. In animations, things are erased or mutated or resurrected; things are not categorized but transformed. Levitt thus argues that animation is "the dominant medium of our time" (2018, 1; and see Manovich 2016).

These discussions of post-cinematic animation are of particular interest to the visualization of urban spaces, because in animated post-cinematic images Cartesian notions of space and time – central to the planner's analytic eye and to cinematic forms of representation – no longer apply. Objects can morph and viewers no longer necessarily inhabit "a kind of ocular-sensory, embodied identity" (Elsaesser 2013a, 32; and see Denson 2016; 2020; Rose 2021). Observers can fly and zoom, and/or be in multiple times and spaces at once. Digital images often suspend the human point of view and human scale (Denson 2016; Elsaesser 2013a, 33); spatial and temporal continuity is "fractured, devalued, fragmented, and reduced to incoherence" (Shaviro 2016, 55). Images are no longer representational but resemble more the artifice of hand-drawn cartoons or paintings (Manovich 2016). All this may also contribute to Boyer's sense that cities are no longer imageable: the overview based on visualizing Cartesian space no longer grips in digitally mediated cities.

None of the chapters in this collection address the most spectacular examples of post-cinematic urban animations, which tend to be Hollywood blockbusters or Netflix series about superheroes, alien invasions, or climate catastrophes, or indeed combinations of all three. In those movies, digitally-created visual special effects often picture extraordinary cities, cities which morph and are folded into one another, cities being overwhelmed by fire or tsunami or meteor strike, city buildings dwarfed by space ships or saturated by apocalyptic rain. These cities are pictured from any and all angles and scales. No longer a single point of view framed by perspectival techniques, the spectator becomes a constantly mobile point of view, decentred, zooming and hovering through an environment that seems to have no frame. Elsaesser describes this unanchored viewing, tracking seamlessly through spaces from the nano to the planetary, as "the default value of digital vision" (2013b, 240), and points to its nondigital precedents in a range of efforts to create convincing three-dimensional films.

Nonetheless, there are clear connections between these post-cinematic movies and the rather more prosaic visualizations discussed in the chapters collected here. The first relates to the *processuality* of post-cinematic images and its effects on what those images look like. Digital images can show spaces differently: less tracing and more sculpting.

Several chapters discuss processes that visualize urban worlds from the digital data extracted from it. Two focus on how various kinds of software work with data about urban environments to produce visions of that environment, visions which align with many of the qualities identified with post-cinematic digital images. The first is Joel McKim's discussion of a number of arts-based projects working with artificial intelligence (AI). McKim begins with a clear exposition of deep learning AI as a process that sculpts new kinds of urban images from other visual data. As he explains, different kinds of neural networks are trained to describe any image, initially on the basis of large numbers of manually tagged images. More recently, such deep learning AI can also generate its own images from that same training. McKim discusses a number of critiques of this sort of imaging. In particular, since machine learning is based on humans describing a large set of images using a delimited set of tags (often the Wordnet database of semantic relations), a number of norms and values are embedded in the AI learning via such datasets from the outset. This is of course an increasingly widespread critique of AI, and critics often focus on the racism, sexism and classism enacted in the tagging process (Benjamin 2019; Noble 2018).

As McKim notes, many criticisms of AI assume that AI are representational, and accuse AI of misrepresenting the actual world because they have been trained wrongly, as it were. However, McKim's chapter also hints at the processual agency of such AI. They may work with images that look as if they are lens-based and they may therefore be said to mis-label what a lens apparently shows – in a cinematic moment. But the three artists' projects discussed in McKim's chapter also have post-cinematic elements. Their sense of presence and stability is not secure. As McKim describes them, and indeed as one is titled, they are uncanny, hallucinatory, sinister. They are both recognizable and not. Their AI picture, or search for, objects that in part have been designed by AI. This is close to how Levitt (2018, 51) sees animations: they are sufficiently recognizable but never entirely so. It is not that they have no relation to the visible world, but rather that that relation is no longer representational.

Another kind of real-but-not digital image of many city streets are images generated by the various sensing technologies embedded in autonomous vehicles. These are images generated by digital devices which show urban environments in ways that have very little in common with film or photography: like McKim's case studies, what they show are recognizable as streets, but point clouds and technicolour skeletons are less familiar. Sam Hind's chapter focusses on the dynamics of AI processing that control how autonomous vehicles navigate streets, particularly busy urban streets, and which generates those images. The main focus of his chapter is the process of "real-time" recognition done by autonomous vehicles' onboard lidar devices (lidar is short for "light detection and ranging"). Hind describes that process as composed of data generation, capture and processing – as per accounts of post-cinematic images – and it is that processing which both generates decisions about the vehicle's mobility and which also, in the process, sees urban spaces in new ways. If McKim only hints at the uncanny newness of machine-learnt imagery, however, Hind argues explicitly that this way of seeing moves quite radically away from familiar ways of seeing, and also from the ocular-sensory, embodied identity of cinematic observer. As Hind notes, autonomous vehicles do not require direct human involvement to engage in their sensemaking activities. Rather, the various technologies continually and processually "broker human accessibility" (Hansen 2015, 6) to the urban environment through new visual forms.

Ayona Datta's chapter explores another technology which could be described as brokering human accessibility to the urban environment: the smartphone with a WhatsApp messaging app. Her analysis focusses more on the human aspects of that brokering. Again, this chapter does not suggest – or does not only suggest – that smartphones allow a more accurate representation of urban life. Rather, Datta emphasizes the ongoing journeys, communications, connections, associations, and interceptions enabled by WhatsApp in the hands of a group of young women in Delhi's urban periphery. By focussing on their everyday and ongoing uses of the messaging app, Datta suggests that complex negotiations over urban space are constantly enacted and re-enacted. They are performed again and again, with and through the use of WhatsApp. She thus indicates how human accessibility to urban spaces is rendered processual when mediated by digital technologies that are themselves processual (see also Rose 2017). WhatsApp written and audio messages, as well as photos and videos, document urban encounters but also co-constitute their users.

Asli Duru also explores what happens as a city is encountered through mobile visualizing devices, in her case GoPro cameras and smartphones. Her account of an Istanbul neighbourhood emerges from her own imbrication in the enactment of digital mediations, as a researcher. She is concerned to evoke "a sense of the existing and emergent worlds, hierarchies and sensitivities that come alive through the interactions between visual practices, things and subjects". This produces a different kind of knowledge about urban spaces which, like the urban geographies of Datta's collaborators, is not revelatory but processed on-the-fly: Duru describes it as "speculative" rather than inscriptive. It interrogates visibility itself through insisting, through the affordances of digital image making and editing, on the provisionality of what is shown.

Several chapters emphasize the multiple temporalities that emerge with the processuality of digital images. The emphasis on processuality and emergence in the chapters by McKim, Hind and Datta tend to focus on the in-the-moment working of software and hardware. The chapter by Scott Rodgers also elaborates a distinctively digital temporality which is experienced as "now". Rodgers explores how the circulation of images on Facebook and Twitter mediates the making of a local area in north London, specifically how the area took shape as a cycling infrastructure scheme was discussed online. He explores how these social media platforms translate asynchronous images and other data into an apparently-real-time experience of immediacy. While social media is experienced in and as the present or the "now" as people scroll through their feeds, what they are seeing might have been uploaded, or commented on, or shared, at many different moments in the past. Rather like the uncanny spatialities discussed by McKim and Hind, Rodgers proposes that the temporalities of online images are paradoxical: "A succession of 'nows'."

Duru also suggests that digital imagery can articulate not only immediate urban experience but also memories of past experiences. Memories infiltrate images too, again rendering them ambiguous. The chapter by Monica Degen and Isobel Ward also explores the multiplicity of temporalities enacted in digital visualizations of urban locations. Their case study is an urban regeneration project in London, and the digital images produced by the many stakeholders in the project. Degen and Ward are particularly interested in the multiple temporalities that are enacted as the plans for the area have developed. They point to a strong sense of the importance of the historic buildings in the area, which has been mobilised to resist past redevelopment plans, and the complex diurnal rhythm of the area's workers and inhabitants through the area's workplaces, clubs, and residences. The latter has been mobilised by one of the project's stakeholders and expressed in the form of large-scale photographic portraits of a diverse range of local residents, while other stakeholders have curated online archives of historical and contemporary images of the area. Still other stakeholders share images of branded cultural events, or picture the future development using computer generated images of its proposed new buildings. While it is the case that cities have very often been seen as palimpsests of different historical moments, Degen and Ward's analyses of digital images of the area's past, present and future suggest that the digital technologies have intensified that multiplicity.

The second aspect of discussions of post-cinematic and post-photographic images that is relevant to several of the chapters here is that digital images circulate. A critical aspect of the sociomaterial affordances of digital images is that they are designed to be distributed digitally (Munster and MacKenzie 2019; Rubinstein and Sluis 2008). The shift from analogue to digital popular photography, for example, was enabled not only by digital cameras but also by increasingly seamless connections between cameras, other viewing devices like computers, and then phones and social media platforms. McQuire (2016) emphasizes the resulting ubiquitous availability of media content as a key aspect of the digital mediation of cities, and accounts of post-cinema pick up on this ubiquity by emphasizing digital movies' ability to be watched on digital screens in all sorts of situations (Casetti 2015). This is one of the ways in which digital imagery must be placed in relation to platform urbanism. Much of the data harvested and distributed by platforms takes visual form on screens. Films and videos, all sorts of photos, animated graphics, memes and gifs, app icons, and more appear on screens that are themselves visual user interfaces. Thinking about the digital visual mediation of urban space must therefore also consider how their patterns of image distribution have consequences.

But the mediation of urban space by digital images shared via social media platforms is a little more complicated than everything being viewable on any screen. Not everything can be shared: technical incompatibilities between devices and software, sometimes generated specifically to protect copyright or a platform, create frictions. Many images are shared as part of an exploitative global division of digital creative labour, through which repetitive processing tasks are sent to cheap labour markets in the Global South (Chung 2018; Murphy and Walker 2019; Rose, Degen, and Melhuish 2014). And on social media, there is evidence that the uneven clustering of likes, follows and comments enacts differentiation between urban spaces.

A study of Instagram use in Amsterdam is instructive here. Boy and Uitermark (2017) analysed 400,000 geotagged Instagram posts from Amsterdam. While they do give some attention to their visual content, they are also particularly interested in the distribution of those posts across different users. They identify what parts of Amsterdam appear most frequently on Instagram and also identify different clusters of users who like and comment on each other's posts; they show that different clusters tend to picture specific parts of the city. Boy and Uitermark thus demonstrate that Instagram mediates Amsterdam not only in terms of how locations are pictured but also by co-constituting different social groupings in relation to those places ("locally oriented gentrifiers", for example, who post a lot of photos with a neighbourhood vibe, or the "vanguard of lifestyle promoters" who post fewer pictures and focus more on their lifestyle aesthetics). What this demonstrates is that it is not just the visual content of Instagram pictures that matters to how social media images shape urban space. So too do the social groupings co-constituted with the patterns of their production and distribution (see also Crang, Crosbie, and Graham 2006). The chapter by Rodgers also focusses directly on the constitution of an urban space – in his example, a local neighbourhood – through the circulation of images on social media platforms, in this case Facebook and Twitter. Rodgers examines how images are embedded in lively debates, banter, and speculation across these platforms, often shared multiple times. These practices produce animated visual environments through various platforms' screen interfaces.

The argument that the extensive distribution of digital images both picture urban spaces (among other urban-related kinds of images) and also create an image-saturated environment is taken in a somewhat different direction by Giorgia Aiello, in her chapter on stock images in urban locations. Aiello's chapter discusses the pervasive use of digital stock photographs with "uplifting visual content" to enliven shopfronts and streets. Aiello explores how stock images are often used not to advertise particular businesses or products (though they often do that on urban billboards and screens) but to create a mood of comforting familiarity in many public spaces. Aiello argues that their ambience is an effect as much of their ubiquity as of their content, making the inhabitants of urban space feel good about it. Her chapter, with Rodgers', thus underlines how it is not only the visual content of digital images that mediates urban space but also their distribution across many surfaces. Furthermore, like the chapters by Rodgers and Degen and Ward, Aiello's also contributes to a body of work which emphasizes how the digital mediation of cities is producing particular urban atmospheres. Degen and Ward, for example, describe how councils and developers increasingly post images on social media that picture the desirability of their projects in terms of what they will feel like, in efforts towards urban placemaking (see also Degen and Rose 2022).

Aiello's discussion of mood and atmosphere also addresses the issue of what it feels like to see digital images in city spaces. This is also the focus of the chapters by Krajina and Rose. Discussions of post-cinematic spectatorship suggest that in relation to the movies at least, sensation and spectacle become more important than narrative or spatial coherence. Just as the post-cinema film pivots on processual transformation and metamorphosis, there is that sense that the viewer too is no longer a stable point but is also assembled and reassembled as they experience post-cinematic affect. Because digital images are the result of the constant processing data by software, there's a sense that the viewing experience is also somehow liquid: viewing becomes more of a live event as sensations of "real-time" feedback become pervasive (McQuire 2016, 5).

Gillian Rose's chapter is an account of the spectating body in digitally mediated cities, which returns to the processuality of the digital image. In that chapter, I explore the implications of the notion of animation, as discussed by Levitt (2018), for viewers of the flows of digital images of and in urban spaces. Particular kinds of bodies coagulate at the interface between digital images and their viewers; flesh is organized visually and spatially. I sketch the ways in which bodies are seen and see in representational visual regimes, as well as in post-cinematic, animatic ways of seeing. I then explore how animated bodies look, and suggest that this has important implications for the bodies doing that looking in the digital image-saturated environments of cities. Like the images that constitute urban code/space, bodies in that space are also constantly emergent, mobile, fluid and mutating.

Zlatan Krajina explores experiences of the mediation of cities by digital images in relation to longstanding conceptualizations of urban space as public space. As he notes, urban public space has very often been understood as constituted by encounters between bodies, speech acts, and objects. What happens then when many of those objects are digital images materialized? His answer further deepens this book's focus on ambiguity, uncanniness, and paradox. Krajina discusses how encounters in urban code/space entail an attentiveness to the city while being otherwise engaged. Through three case studies, he elaborates different configurations of the intertwining of attention and distraction. In all three, he evokes particular tones and moods that are both expressive and diffused.

All of these chapters explore different aspects of the digital processing of images and its configuration of urban spatialities and temporalities. There are two more themes threading through these chapters. One is social power. All the chapters affirm that many kinds of power are imbricated in digital images. Sometimes this power remains representational: digital images of cities represent only some kinds of city spaces and bodies, only some memories and futures. In other examples, power settles in the capacity to move or not, to be mobile or not through various spaces both material and digital. In other chapters, power is in the constitution of certain kinds of bodies, temporalities, and spatialities, in urban code-space. These forms of power have different modalities, but the chapters by Datta, Hind, and Duru in particular all propose that, diffuse and atmospheric as processual as digital images may be, they nonetheless can be imbricated in questions of bodily violence and even death.

The final contribution made by the authors of the chapters collected here relates to the research methods most appropriate to how cities are seen digitally. Of the chapters gathered here, Duru's reflects at greatest length on the methodological implications of digital ways of seeing urban spaces. Resisting the modes of visualizing cities which replicate the analytic eye of representationalist images, she counter proposes a speculative research methodology as itself an appropriately processual approach to urban code/space. Rodgers experiments with data visualization methods that present images en masse, and Datta has developed an online multimedia website which works with a number of different visuals generated as part of the collaboration; the site is interactive and offers multiple encounters to its visitors rather than a single urban reality. Duru and Aiello immerse themselves in the cities and theorize from their own embodied experiencing of visual atmospheric code/space, a method which might align with the focus of Rose's chapter. All this suggests that new ways of seeing urban space digitally also require (some) new methods of researching urban space visually which more closely align with new, post-cinematic ways of seeing. This is an aspect of the arguments presented here which deserves considerable further elaboration and experimentation.

Conclusion

This chapter builds on the arguments made in the various chapters gathered in this collection, as well as from a rich body of work on digital images, urban screens and post-cinema. There are without doubt other relevant bodies of work: on sensory urban atmospheres (Sumartojo and Pink 2019; Degen and Rose 2022) and nonrepresentational urbanism (Thrift 2014), for example; as well as the posthumanist, technosocial and materialist theories assumed by many of the chapters here but not unpacked. And important forms of visualizing cities are absent in this collection: as the introduction has already noted, there are no chapters on the cities to be found in superhero movies, nor on computer games or influencer feeds; there is no discussion of digital visual surveillance in cities (Zuboff 2019).

Nevertheless, the rich discussions here do clearly demonstrate that digitally mediated cities are visualized no less thoroughly than cities were before digital technologies became so commonplace. Cities have not become any less visible since the heyday of modernist planning. Rather, what has changed is the form of the visualizing that brings them into (new forms of) visibility. Planning and much urban management in the nineteenth and twentieth centuries relied on a representationalist visual paradigm that valorised particular kinds of images, which were assumed to display a particularly close relation to urban reality. In this they were typical of a visual culture in which photography and film were also broadly taken to be representational of what they pictured. This representationalism depended on both some of the affordances of lens-based technologies and on the specific ways in which they were interpreted. This paradigm weakened towards the end of the twentieth century. New digital visualizing technologies and new ways of making images with them, as well as new forms of urban governance, became increasingly pervasive. Now, as we enter the third decade of the twenty-first century, it is more evident that many cities remain as visible as they ever were: indeed, in an era of "ubiquitous photography", it could be claimed that cities are more visible than ever (Hand 2012). What has changed, though, is the kinds of imagery through which cities become seen. No longer necessarily offering analytical insight into the truth of urban life, digital images are *processual* and *circulatory*, and, as this introduction has discussed, this has implications for how images organize urban spatialities and temporalities.

Finally, it is interesting to note that all of these chapters have a somewhat oblique relationship to "the image". While all focus on particular images, often many kinds of images, few spend time interpreting specific images. There is little of the careful decoding of specific images using the conventional critical toolkit of semiology or discourse analysis. This reflects the particular kind of visuals these chapters address. This introduction has used a variety of terms to refer to these digital images: post-cinematic, post-photographic, animations, digital images, digital visualizations. All of these terms emphasize that the images which mediate urban code/space are digitally processed and ubiquitously distributed. These aspects of their digitality produce not only the uncanny or paradoxical spatial and temporal effects of their visual content, as explored by several chapters here. It also means that these images are multiple. They can be made and remade, they are refreshed and renewed, and they travel through networks, servers and screens to appear many times on many interfaces. As the chapters here suggest, this means that they have to be approached less as single objects and more as environments or atmospheres, which are visual but do not render cities imageable in the ways cities have been for the past two or three hundred years. What the chapters in this book suggest is that this does not lead understanding how cities are rendered visible into a blind spot: rather, it encourages researchers to see cities differently both when we are in urban spaces but also when we design our investigations into those spaces. Picturing the city representationally no longer quite works. But the city as a system of systems, as an ecology of decentred and recombinatory platforms (Barns 2020a), or as a multispeed city with variable geometries (Crang, Crosbie, and Graham 2006), can certainly be visualized by digital images that themselves flow and morph. Cities are thus constantly transformed in the "mixed-space effect" of animation (Levitt 2018, 68). And those pictures, digital all the way down, also in part constitute a distinctively digital urban geography.

References

Anthopoulos, Leonidas. 2017. "Smart Utopia vs Smart Reality: Learning by Experience from 10 Smart City Cases." *Cities* 63 (March): 128–48.

Ash, James, Ben Anderson, Rachel Gordon, and Paul Langley. 2017. "Unit, Vibration, Tone: A Post-Phenomenological Method for Researching Digital Interfaces." *Cultural Geographies* 25 (1): 165–81.

Aurigi, Alessandro, and Katherine Willis, eds. 2020. *The Routledge Companion to Smart Cities.* Abingdon: Routledge.

Barns, Sarah. 2020a. *Platform Urbanism: Negotiating Platform Ecosystems in Connected Cities.* Basingstoke: Palgrave MacmIllan.

———. 2020b. "Tools of the Trade: Urban Planning, Urban Media and the Refashioning of Urban Space." In *The Routledge Companion to Urban Media and Communication*, edited by Zlatan Krajina and Deborah Stevenson, 236–47. Abingdon: Routledge.

Batty, M. 1990. "Intelligent Cities: Using Information Networks to Gain Competitive Advantage." *Environment and Planning B: Planning and Design* 17 (3): 247–56.

Benjamin, Ruha. 2019. *Race After Technology: Abolitionist Tools for the New Jim Code.* Cambridge: Polity Press.

Besse, Jean-Marc. 2013. "European Cities from a Bird's-Eye View: The Case of Alfred Guesdon." In *Seeing From Above: The Aerial View in Visual Culture*, edited by Mark Dorrian and Frédéric Pousin, 66–82. London: IB Tauris.

Boy, John D, and Justus Uitermark. 2017. "Reassembling the City through Instagram." *Transactions of the Institute of British Geographers* 42 (4): 612–24.

Boyer, M. Christine. 1994. *The City of Collective Memory: Its Historical Imagery and Architectural Entertainments*. London: MIT Press.

Boyer, Marie-Christine. 1996. *CyberCities: Visual Perception in the Age of Electronic Communication*. New York: Princeton Architectural Press.

Browne, Simone. 2015. *Dark Matters: On the Surveillance of Blackness*. London: Duke University Press.

Cardullo, Paolo, and Rob Kitchin. 2019. "Smart Urbanism and Smart Citizenship: The Neoliberal Logic of 'Citizen-Focused' Smart Cities in Europe." *Environment and Planning C: Politics and Space* 37 (5): 813–30.

Casetti, Francesco. 2015. *The Lumière Galaxy: Seven Key Words for the Cinema to Come*. New York: Columbia University Press.

Castells, Manuel. 1989. *The Information City: Information Technology, Economic Restructuring, and The Urban-Regional Process*. Oxford: Blackwell.

Chung, Hye Jean. 2018. *Media Heterotopias: Digital Effects and Material Labour in Global Film Production*. London: Duke University Press.

Clark, Catherine E. 2018. *Paris and The Cliche of History: The City and Photographs, 1860-1970*. Oxford: Oxford University Press.

Cowley, Robert, Simon Joss, and Youri Dayot. 2018. "The Smart City and Its Publics: Insights from across Six UK Cities." *Urban Research & Practice* 11 (1): 53–77.

Crang, Michael, Tracey Crosbie, and Stephen Graham. 2006. "Variable Geometries of Connection: Urban Digital Divides and the Uses of Information Technology." *Urban Studies* 43 (13): 2551–70.

Crary, Jonathan. 1990. *Techniques of the Observer: On Vision and Modernity in the Nineteenth Century*. Cambridge, MA: MIT Press.

Degen, Monica, and Gillian Rose. 2022. *The New Urban Aesthetic: Digital Experiences of Urban Change*. London: Bloomsbury Academic.

Denson, Shane. 2016. "Crazy Cameras, Discorrelated Images, and the Post-Perceptual Mediation of Post-Cinematic Affect." In *Post-Cinema: Theorizing 21st-Century Film*, edited by Shane Denson and Julia Leyda, 193–233. Falmer: Reframe.

———. 2020. *Discorrelated Images*. Durham, NC: Duke University Press.

Denson, Shane, and Julia Leyda, eds. 2016. *Post-Cinema: Theorizing 21st-Century Film*. Falmer: Reframe. http://reframe.sussex.ac.uk/post-cinema/.

Donald, James. 1999. *Imagining the Modern City*. London: Athlone Press.

Elsaesser, Thomas. 2013a. "Digital Cinema: Convergence or Contradiction?" In *The Oxford Handbook of Sound and Image in Digital Media*, edited by Carol Vernallis, Amy Herzog, and John Richardson, 12–44. Oxford: Oxford University Press.

———. 2013b. "The 'Return' of 3-D: On Some of the Logics and Genealogies of the Image in the Twenty-First Century." *Critical Inquiry* 39 (2): 217–46.

Eubanks, Virginia. 2017. *Automating Inequality: How High-Tech Tools Profile, Police, and Punish the Poor*. New York: St Martin's Press.

Flint, Kate. 2000. *The Victorians and the Visual Imagination.* Cambridge: Cambridge University Press.

Gordon, Eric. 2010. *The Urban Spectator: American Concept Cities from Kodak to Google.* 1st ed. Hanover, NH: Dartmouth College Press.

Graham, Stephen D. N. 2005. "Software-Sorted Geographies." *Progress in Human Geography* 29 (5): 562–80.

Halpern, Orit. 2015. *Beautiful Data: A History of Vision and Reason since 1945.* Durham: Duke University Press.

Hand, Martin. 2012. *Ubiquitous Photography.* Cambridge: Polity Press.

Hansen, Mark B. N. 2015. *Feed-Forward: On the Future of Twenty-First-Century Media.* Chicago: University of Chicago Press.

Hodson, Mike, Julia Kasmire, Andrew McKeekin, John G Stehlin, and Kevin Ward, eds. 2020. *Urban Platforms and the Future City: Transformations in Infrastructure, Governance, Knowledge and Everyday Life.* Abingdon: Routledge.

Hollands, Robert G. 2008. "Will the Real Smart City Please Stand Up?" *City* 12 (3): 303–20.

Huyssen, Andreas. 2008. "Introduction: World Culture, World Cities." In *Other Cities, Other Worlds: Urban Imaginaries in a Globalizing Age,* edited by Andreas Huyssen, 1–23. Durham: Duke University Press.

Kember, Sarah, and Joanna Zylinska. 2012. *Life After New Media: Mediation as a Vital Process.* Cambridge, MA: MIT Press.

Kitchin, Rob, and Martin Dodge. 2011. *Code/Space: Software and Everyday Life.* Cambridge, MA: MIT Press.

Leszczynski, Agnieszka. 2019. "Platform Affects of Geolocation." *Geoforum* 107: 207–15.

Levitt, Deborah. 2018. *The Animatic Apparatus: Animation, Vitality, and the Futures of the Image.* Winchester: Zero Books.

Lindner, Christoph, and Miriam Meissner, eds. 2019. *The Routledge Companion to Urban Imaginaries.* Abingdon: Routledge.

Luque-Ayala, Andrés, and Simon Marvin. 2016. "The Maintenance of Urban Circulation: An Operational Logic of Infrastructural Control." *Environment and Planning D: Society and Space* 34 (2): 191–208.

Manovich, Lev. 2006. "The Poetics of Augmented Space." *Visual Communication* 5 (2): 219–40.

Manovich, Lev. 2016. "What Is Digital Cinema?" In *Post-Cinema: Theorizing 21st-Century Film,* edited by Shane Denson and Julia Leyda, 20–50. Falmer: Reframe. http://reframe.sussex.ac.uk/post-cinema/.

Marvin, Simon, Andres Luque-Ayala, and Colin McFarlane, eds. 2016. *Smart Urbanism: Utopian Vision or False Dawn?* Abingdon: Routledge.

McQuire, Scott. 2016. *Geomedia: Networked Cities and the Future of Public Space.* Cambridge: Polity Press.

———. 2020. "An Archeology of the Media City: Toward a Critical Cultural History of Mediated Urbanism." In *The Routledge Companion to Urban Media and Communication*, edited by Zlatan Krajina and Deborah Stevenson, 13–22. Abingdon: Routledge.

Mitchell, W. J. T. 1992. *The Reconfigured Eye: Visual Truth in the Post-Photographic Era*. Cambridge, MA: MIT Press.

Mitchell, William J. 2003. *Me++: The Cyborg Self and the Networked City*. Cambridge, MA: MIT Press.

Munster, Anna, and Adrian MacKenzie. 2019. "Platform Seeing: Image Ensembles and Their Invisualities." *Theory, Culture & Society* 36 (5): 3–22.

Murphy, Ian, and Saint John Walker. 2019. "'Make It More Invisible': Hollywood VFX Training Crosstalk on the Transatlantic Fiber Optic." *Animation Studies* 15. https://journal.animationstudies.org/ian-murphy-saint-john-walker-make-it-more-invisible-hollywood-vfx-training-crosstalk-on-the-transatlantic-fiber-optic/.

Noble, Safiya Umoja. 2018. *Algorithms of Oppression: How Search Engines Reinforce Racism*. New York: New York University Press.

Otter, Chris. 2008. *The Victorian Eye: A Political History of Light and Vision in Britain, 1800-1910*. London: University of Chicago Press.

Packer, Jeremy, and Stephen B. Grofts Wiley. 2012. "Introduction: The Materiality of Communication." In *Communication Matters: Materialist Approaches to Media, Mobility and Networks*, edited by Jeremy Packer and Stephen B. Crofts Wiley, 3–16. London: Routledge.

Parks, Lisa, and Nicole Starosielski. 2015. "Introduction." In *Signal Traffic: Critical Studies of Media Infrastructures*, edited by Lisa Parks and Nicole Starosielski, 1–27. Urbana: University of Illinois Press.

Pinney, Christopher, and Nicolas Peterson, eds. 2003. *Photography's Other Histories*. London: Duke University Press.

Prakash, Gyan. 2008. "Introduction." In *The Spaces of the Modern City: Imaginaries, Politics, and Everyday Life*, edited by Gyan Prakash and Kevin Michael Kruse, 1–18. Oxford: Princeton University Press.

Ristola, Jacqueline. 2017. "Realist Film Theory and Flowers of Evil: Exploring the Philosophical Possibilities of Rotoscoped Animation." *Animation Studies* 12. https://journal.animationstudies.org/category/maureen-furniss-award-2017/jacqueline-ristola-realist-film-theory-and-flowers-of-evil-exploring-the-philosophical-possibilities-of-rotoscoped-animation-winner/.

Rose, Gillian. 2010. *Doing Family Photography: The Domestic, The Public and The Politics of Sentiment*. Farnham: Ashgate.

———. 2017. "Posthuman Agency in the Digitally Mediated City: Exteriorization, Individuation, Reinvention." *Annals of the American Association of Geographers* 107 (4): 779–93.

———. 2018. "Look InsideTM: Visualising the Smart City." In *An Introduction to Geomedia: Spaces and Mobilities in Mediatized Worlds*, edited by Karin Fast, André Jansson, Johan Lindell, Linda Ryan Bengtsson, and Mekonnen Tesfahuney, 97–113. London: Routledge.

Rose, Gillian, Monica Degen, and Clare Melhuish. 2014. "Networks, Interfaces, and Computer-Generated Images: Learning from Digital Visualisations of Urban Redevelopment Projects." *Environment and Planning D: Society and Space* 32 (3): 386–403.

Rose, Gillian, Parvati Raghuram, Sophie Watson, and Edward Wigley. 2020. "Platform Urbanism, Smartphone Applications and Valuing Data in a Smart City." *Transactions of the Institute of British Geographers* 46 (1):59–72.

Rose, Gillian, and Divya Tolia-Kelly, eds. 2012. *Visuality/Materiality: Images, Objects and Practices*. Farnham: Ashgate.

Rubinstein, Daniel, and Katrina Sluis. 2008. "A Life More Photographic: Mapping the Networked Image." *Photographies* 1 (1): 9–28.

Sadowski, Jathan. 2020. "Cyberspace and Cityscapes: On the Emergence of Platform Urbanism." *Urban Geography*, January, 1–5.

Shaviro, Steve. 2016. "Post-Continuity: An Introduction." In *Post-Cinema: Theorizing 21st-Century Film*, edited by Shane Denson and Julia Leyda, 52–64. Falmer: Reframe. http://reframe.sussex.ac.uk/post-cinema/.

Srnicek, Nick. 2014. "Auxiliary Organs: An Ethics of the Extended Mind." In *The Imaginary App*, edited by Paul D Miller and Svitlana Matviyenko, 71–82. London: MIT Press.

Stacey, Noni. 2020. *Photography of Protest and Community: The Radical Collectives of the 1970s*. London: Lund Humphries.

Sumartojo, Shanti, and Sarah Pink. 2019. *Atmospheres and The Experiential World: Theory and Methods*. London: Routledge.

Tagg, John. 1988. *The Burden of Representation: Essays on Photographies and Histories*. Basingstoke: Macmillan Education.

Thrift, Nigel. 2014. "The 'Sentient' City and What It May Portend." *Big Data & Society* 1 (1): 1–21.

Uricchio, William. 2011. "The Algorithmic Turn: Photosynth, Augmented Reality and the Changing Implications of the Image." *Visual Studies* 26 (1): 25–35.

Willis, Katherine, and Alessandro Aurigi. 2018. *Digital and Smart Cities*. Abingdon: Routledge

Wilmott, Clancy. 2016. "In-between Mobile Maps and Media: Movement." *Television & New Media* 18 (4): 320–35.

Zuboff, Shoshana. 2019. *The Age of Surveillance Capitalism: The Fight for A Human Future at the New Frontier of Power*. London: Profile.

About the author

Gillian Rose is Professor of Human Geography at the University of Oxford and a Fellow of the British Academy and of the Academy of Social Sciences. She is the author of *Feminism and Geography* (Polity, 1993), *Doing Family Photography* (Ashgate, 2010), *The New Urban Aesthetic: Digital Experiences of Urban Change* written with Monica Degen (Bloomsbury, 2022) and *Visual Methodologies* (Sage, fifth edition 2022), as well as many papers on images, visualising technologies and ways of seeing in urban, domestic and archival spaces. Her current research interests focus on contemporary digital visual culture.

2. Deep Learning the City: The Spatial Imaginaries of AI

Joel McKim

Abstract
This chapter examines how deep learning neural networks and computer vision technologies are impacting the design, organization and occupation of cities. It begins by providing a brief history of the development of "deep learning" approaches to artificial intelligence. The chapter then focuses on the ways artists and designers have begun to engage with deep learning and computer vision in order to highlight critical questions, especially about the ethical issues surrounding the training datasets these systems depend on. The chapter discusses three art and design examples that shift focus specifically towards the city and spatial concerns, considering the ways these works explore machine learning (the opportunities it presents and the problems it raises) within a specifically architectural or urban context.

Keywords: deep learning, artificial intelligence, art, design, architecture

In the summer of 2019, the subterranean boiler room of New York's popular Chelsea Market opened to the public for the first time. Transformed from its original use, the room was now an art space, run by ARTECHOUSE, a self-described digital art organization dedicated to experiments in art and technology with exhibition venues in Washington and Miami, in addition to New York. For the inaugural exhibition in the 6,000 square foot boiler room, visitors were invited to enter a *Machine Hallucination* designed by Turkish-born artist Refik Anadol. Showing off the space's sophisticated projection technology, Anadol's immersive installation covered virtually every surface of the room in the kind of inexact, morphing images we've come to associate with artificial intelligence. Anadol's work is indeed an

Rose, G. (ed.), *Seeing the City Digitally. Processing Urban Space and Time.*
Taylor & Francis Group, 2022
DOI 10.5117/9789463727037_CH02

experiment in AI-generated images – the artist has been working with AI and machine learning since completing a residency at Google's Artists and Machine Intelligence Program in 2016. The *Machine Hallucination* installation seems doubly relevant to a chapter on deep learning technologies and the city – the work is itself architectural, enveloping an interior space in a surround of AI-generated visuals, and those images, however dream-like or vague, are also recognizably urban. Anadol trained his machine-learning system on 100 million photographs of New York City found on social networks, effectively teaching it to produce its own images of the city based on this archive of public memories.

In some ways Anadol's work is representative of a growing number of artists and designers employing AI technologies such as machine learning in their work, sometimes as methods of aesthetic experimentation and other times as a means of questioning the social, political and economic impact of these fast-developing technologies. Understandably, the human form, and the human face in particular, has featured prominently in many of these works, with artists creating new AI-generated forms of portraiture or producing critical design projects examining the implications of machine-learning powered systems of facial recognition or human classification. Anadol's more unusual focus on images of the built environment invites a consideration of how these technologies are being deployed in the areas of architecture and city planning, but also how artists and designers are creating works that explore questions of AI and urban space. This chapter will outline some of the ways artists and designers are working with deep learning technologies, while highlighting works that address the images and spaces of the city specifically. At the risk of exhausting the limits of both my own technical knowledge and the patience of my readers, the chapter will begin by providing a brief history of the development of "deep learning" approaches to AI. While the computational and mathematical details of AI and machine learning systems can be difficult to summarize effectively or succinctly, I believe it's becoming increasingly important for scholars of the arts and humanities to attempt to engage with these systems at a technical level. As these technologies become central to contemporary visual culture, we need to develop a better understanding of the computational infrastructures that are producing a growing number of the cultural objects and images that surround us (from notorious "deep fake" videos, to AI "up-scaled" video games, to algorithmically filtered photographs).[1] After providing an

1 A number of arts and humanities-based research projects are beginning to take on the task of mapping the aesthetic and cultural significance of new developments in machine imaging

introduction to this technical context and terminology, the chapter will then outline some of the ways artists and designers have begun making use of deep learning technologies, highlighting the critical questions that have surfaced in this work. The ethical issues surrounding the training datasets these systems depend on emerges as a recurrent theme. Finally, the chapter will discuss three art and design examples that shift focus specifically towards the city and spatial concerns, considering the ways these works explore machine learning (the opportunities it presents and the problems it raises) within a specifically architectural or urban context, namely: the *Uncanny Rd.* online generative tool, Simone C. Niquille's CGI-based film *Homeschool*, and a trio of Forensic Architecture investigations: *Triple-Chaser*, *The Battle of Ilovaisk*, and *Model Zoo*.

A very brief history of deep learning

Deep learning (a term that now circulates frequently, but often without a great deal of explanation) is a specific approach to artificial intelligence and machine learning that involves a method based on a hierarchy of concepts. The fundamental idea being that a machine can learn more complex concepts by building on simpler concepts. As a result, the approach usually involves the use of multi-layered, and therefore "deep", artificial neural networks. We could imagine, for instance, a neural network trained to recognize hand-written numbers, a frequent example used in introductions to machine learning (see Nielsen 2019 and Bishop 2006). Early layers of the network might recognize very simple forms like edges, feeding this information forward to subsequent layers that recognize increasingly complex patterns (like loops or intersecting lines), until an output layer eventually recognizes the form of the numbers themselves. As Goodfellow, Bengio and Courville outline in their 2016 textbook on the subject, deep learning is a solution to the problem of machine learning that has a long history with an ebb and flow of acceptance within the field of AI research. They date the emergence of the concept of deep learning as far back as the 1940s, with its current resurgence as the dominant paradigm of AI beginning in 2006 (12). Three waves of development during this quite long history are identified by the

and computer vision, including the Machine Vision in Everyday Life research project led by Jill Walker Rettberg at the University of Bergen, the Operational Images and Visual Culture project led by Jussi Parikka at FAMU in Prague, and my own Pre-Histories of Machine Vision research project conducted at the V&A Museum.

authors: a cybernetic moment in the 1940s through 1960s that eventually wanes; a return to the concept through notions of "connectionism" in the 1980s and 1990s; and the current period spurred on by the breakthroughs of contemporary computer scientists like Geoffrey Hinton. I'll attempt to provide here a very rough sketch of the development of deep learning across these three waves or periods.

The first cybernetic moments of deep learning research emerged from early neural network research and an interest in models of biological brain function that were developing at the time. In 1943 the neuroscientist and cybernetician Warren McCulloch and the logician Walter Pitts proposed the first computational model of a neural net comprised of individual, largely undifferentiated neurons (the basic working unit of the brain, processing and transmitting cellular signals). Inspired by the extremely influential ideas of information theory being formulated by both Claude Shannon and Norbert Wiener at the time, McCulloch and Pitts proposed that the biological system of information exchange that is the nervous system could find analogous form in the logic processing of mathematics. We can view this as the beginnings of a long tradition of conceiving of the human brain as essentially a computation machine and therefore comparable to the digital computers just beginning to emerge at the time. McCulloch and Pitts begin their 1943 paper "A Logical Calculus of the Ideas Immanent in Nervous Activity" with the claim, "Because of the 'all-or-none' character of nervous activity, neural events and the relations among them can be treated by means of propositional logic" (McCulloch & Pitts 1943). In other words, the McCulloch-Pitts neuron, the basic unit of their model, was conceived as kind of logic gate – a linear mathematical function capable of taking a series of weighted inputs and aggregating them to produce a single output or decision. This essential premise – that an artificial neural network is made of a network of connected neurons, each one a mathematical function processing inputs according to varying weights – remains the foundation of contemporary deep learning.

The McCulloch-Pitts neuron would become the inspiration point for artificial neurons to follow, most notably the perceptron algorithm produced by the psychologist Frank Rosenblatt in 1958. Rosenblatt developed the perceptron at the Cornell Aeronautical Laboratory, funded by the US Office of Naval Research (ONR). Although first implemented as software running on an IBM mainframe computer, Rosenblatt intended for the perceptron to be realized as a custom-built machine, a goal which eventually materialized in the form of the "Mark I Perceptron" in the early 1960s. Image recognition was a central task for neural networks from the outset and the first use of the perceptron involved connecting the machine to a simple camera system in which a lighted object was registered by a 20 × 20 array of cadmium sulphide photocells, producing a primitive 400 pixel image (Bishop 2006, 196). The photocells were wired to the neurons of the perceptron at random, demonstrating the system's ability to learn independently. An important distinguishing point from the McCulloch-Pitts neuron was the perceptron's ability to adjust the weighted values of the inputs automatically, rather than by human operator. Rosenblatt's research generated considerable public attention, but he was also considered to be prone to overclaiming, issuing "steady and extravagant statements about the performance of his machine" (McCorduck 2004, 105). After listening to Rosenblatt's initial 1958 press conference for the perceptron, *The New York Times* gushed: "The Navy revealed the embryo of an electronic computer today that it expects will be able to walk, talk, see, write, reproduce itself and be conscious of existence" (in Olazaran 1996 621).

Marvin Minsky was one important figure in the AI community irritated by Rosenblatt's bluster. The two scientists had attended the same high school in the Bronx and had maintained a rivalry throughout their careers (McCorduck 2004, 106). Minsky and Seymour Papert's 1968 book *Perceptrons: An Introduction to Computational Geometry* exposed some of the perceptron's limitations in relation to pattern recognition, classification, and its ability to internally represent its own act of perception. Minksy and Papert who favoured the rival "symbolic" approach to AI have more recently been accused of falsely characterizing the abilities of neural networks, focusing exclusively on the limitations of a single layer perceptron rather than the potential of multi-layered neural networks (which was already evident at the time). While suggesting the story is actually a more complicated one, Mikel Olazaran acknowledges, "according to the official history of the controversy, after Minsky and Papert's study, the neural-net approach was rejected and abandoned" (1996, 640).

According to Goodfellow, Bengio and Courville, the resurgence of interest in neural networks in the 1980s emerged out of the interdisciplinary field of cognitive science through a movement called "connectionism", which rekindled the notion that an interactive network of simple computational units was capable of generating intelligent behaviour (2016, 16). Many of the algorithms still in use in the machine learning of today were developed or optimized during this period. As Adrian Mackenzie notes in his book *Machine Learners*, "the algorithms such as back-propagation used in neural nets have not [...] been radically transformed in their core operations since the 1980s, and even then the algorithms (principally gradient descent) were not new" (2017, 191). Put very simply, back-propagation is an algorithm by which a neural network is capable of optimizing the internal weights of the functions operating in its neurons, a key process in its ability to calibrate and learn. These algorithmic advances included breakthroughs in the field of computer vision and image processing, such as the development of convolutional neural networks – a class of deep learning network still considered to be the most effective for image recognition and classification (Fukushima 1980, LeCun et al. 1999). Convolutional neural networks were inspired by biological visual cortex systems and the sensory processing experiments of the neurophysiologists Hubel and Wiesel (1959). To again simplify greatly, a convolutional neural network employs operations of sub-sampling, filtering and synthesizing (a process of "convolving") in order to optimize its ability to recognize patterns in images.

Goodfellow, Bengio and Courville date the contemporary moment of deep learning to 2006 when significant breakthroughs in the effectiveness and efficiency of neural networks begin to emerge. Given that the basic premise of artificial neural networks and even some of the algorithms still in use date back to the mid-twentieth century, it seems fair to ask what brought about this relatively recent explosion of deep learning development. Most accounts of the growth of the field highlight two factors: the acceleration of computational processing power that has made feasible increasingly large or more efficient neural networks made of multiple layers (sometimes over a hundred), and the availability of large, often tagged, data sets used to train these neural networks. The availability of these large data sets has been fuelled in part by the expanded circulation and archiving of media online. Convolutional neural networks, for example, require images to learn and lots of them. The mass posting of photographs online that has occurred over the past two decades provides an ideal training resource for these networks.

An important example of these two factors coming together (increased processing power and the availability of large training sets) was the development of the convolutional neural network AlexNet, designed by Alex Krizhevsky at the University of Toronto, and published with Ilya Sutskever and Geoffrey Hinton (2012). AlexNet competed in the 2012 ImageNet Large-Scale Visual Recognition Challenge (ILSVRC) that has been a benchmark for computer vision developments. The competition called on research teams to use their deep learning neural networks to classify images from the ImageNet dataset. ImageNet, arguably the most significant computer vision training set, was first unveiled in 2009 by a team of AI researchers at Stanford and Princeton led by Professor Fei-Fei Li, who once described the project as an attempt "to map out the entire world of objects" (in Gershgorn 2017). The dataset now consists of more than 14 million images, often scraped from online photo-sharing sites like Flickr and tagged by a crowdsourced army of workers into over 20,000 categories. These manually labelled datasets are often called "ground truth data" within discourses of deep learning (Schmidt 2019). AlexNet, employing an eight-layer convolutional neural network powered by two graphics processing units (GPUs), outperformed its competitors on its ability to correctly recognize or classify images in the ImageNet collection (images ranging from "container ships" to "Siamese cats"), achieving an impressively low error rate (Wei 2019).

A recent deep learning breakthrough has led to neural networks capable of not only classifying images, but also creating them, a development that has probably played the largest role in bringing wider technical advances in deep learning to public attention. Ian Goodfellow and colleagues invented Generative Adversarial Networks (GANs) in 2014 while Goodfellow was a student at the Université de Montréal under the supervision of Yoshua Bengio (Goodfellow et al. 2014). GANs involve engaging two neural networks (trained on the same dataset) in a kind of recognition game. A "generative network" produces images intended to pass as "candidates" for the dataset. A second "discriminative network" evaluates these generated images, determining how likely they are to be "real" images from the set. Through the learning mechanism of backpropagation both the generative network and the discriminative network gradually become better at their roles in this computational game of fool or be fooled. GANs are largely responsible for producing what has become the popular visual culture of AI, helping to create everything from the infamous "deep fakes" circulating online (the disturbingly iconic Jennifer Lawrence/Steve Buscemi mash-up video, for example) to the AI-generated *Portrait of Edmond Belamy* produced by the Paris-based Obvious collective that caused a media stir when it sold at auction for $432,000 in the Autumn of 2018. It's GAN technology, incidentally, that powered the machine hallucinations of New York City displayed by Refik Anadol in the Chelsea Market boiler room in 2019.

Deep learning in art and design

The art and design projects that make use of deep learning technologies, and GANs most frequently, are often both exploratory and critical in nature. Through initiatives such as the website *This Person Does Not Exist*, even AI industry insiders like Philip Wang (a software engineer at Uber at the time of the site's creation) strike a cautionary note regarding the deceptive potential of the technology. True to its name, *This Person Does Not Exist* generates extremely convincing photorealistic images of otherwise non-existent people. It does so by employing StyleGAN, a generative adversarial network designed by engineers at Nvidia, the leading producer of the GPUs that provide the processing power for most neural networks. Each refresh of the page produces yet another person that does not exist. In describing his motivation for creating the site Wang explains, "I just hope my demonstration raises awareness. Those who are unaware are most vulnerable to this technology" (in Paez 2019). Although despite this expression of concern, Wang also sees positive potential for deep learning technologies and accompanies each image with the tag line "Don't panic. Learn how it works", along with links to YouTube videos explaining the technical details of the GAN-powered human face synthesis algorithms at work in creating these images.[2]

Many of the art and design projects exploring the growth of deep learning and computer vision technologies question the role played by the training sets and hierarchies of classification that serve as the underlying infrastructure of these systems. Adam Harvey's ongoing *MegaPixels* initiative is a good example of this critical perspective, a project described by the artist and researcher as an investigation of "the ethics, origins, and individual privacy implications of face recognition image datasets and their role in the expansion of biometric surveillance technologies" (Harvey 2019). Harvey questions the political implications of datasets such as MegaFace, a training set of 4,753,320 faces derived from public Flickr photo albums, analysing the metadata connected to the dataset's images and revealing the potential violation of Creative Commons licenses involved in their use.

As mentioned in our short history of deep learning, ImageNet is the training set that has almost certainly contributed to recent developments in computer vision and machine learning more than any other. Unsurprisingly, the influential dataset has also been the focus of a number of critical art and design projects. The artist Trevor Paglen has placed ImageNet at the centre of two recent projects, his 2019 Barbican exhibition featured a newly commissioned work entitled *From "Apple" to "Anamoly"* – an array

2 The synthesized faces generated by contemporary GAN technology reproduce some of the desires and anxieties provoked by earlier iterations of computational technologies. These faces recall *Time* magazine's controversial cover for its special fall 1993 issue on immigration and multiculturalism, "The New Face of America", featuring the image of a woman purportedly produced by morphing together the facial features of multiple racial and ethnic groups. Donna Haraway critiques the eliding of messy biological and political difference represented by this technologically composited, universal "SimEve" (1995). We might question what new technophilic fantasies of identity accompany the endless non-faces produced by neural networks.

of approximately 30,000 printed photographs, all images derived from the training set, that virtually covered the sweeping wall of the Curve gallery. Paglen's work displays images belonging to a cross-section of ImageNet's categories, beginning rather innocently with images clustered around the labels "apple", "apple tree" and "fruit", before moving on to more complex and contentious examples, from "minibar" to "abattoir" to "divorce lawyer". As I wrote in a recent review of the exhibition: "It's when people first appear among the photographs of objects that we begin to realize how strange and troubling this exercise in image classification really is. The category 'picker' includes smiling recreational strawberry pickers alongside Indian tea-leaf pickers and impoverished children picking through waste in a landfill. Any contextual distinction between these images is apparently flattened out in the eyes of the machine" (McKim 2019). The biases and absurdities of ImageNet's structure of classification become obvious when we notice, for example, that the labels "investor", "entrepreneur" and "venture capitalist' present almost exclusively images of white, middle-aged men, whereas less flattering categories such as "selfish person", "moneygrubber" and "convict" are considerably more diverse.

As the artist and programmer Nicholas Malevé has pointed out, the classification system for ImageNet is reliant on the WordNet database of semantic relations developed at Princeton: "Pressing into service an existing classification system however brings in its own share of problems, omissions and decision-making issues. WordNet for instance unreflexively integrates and naturalizes racial and gender binaries and its structure contributes to reifying social norms" (2019). The potential problems associated with ImageNet's system of classification were further highlighted in Paglen's *ImageNet Roulette*, a project featured in the "Training Humans" exhibition at the Osservatorio Fondazione Prada that Paglen and Kate Crawford co-curated in 2019. *ImageNet Roulette* is a computer vision system that captures the video image of gallery visitors and assigns them labels from the ImageNet's people categories (an online version of the work was also made available). The labels are often uncomplimentary, gendered and even racist, which Paglen and Crawford defend as a provocation to question the inherent prejudices of the ImageNet dataset and these forms of human categorisation more generally. In their "Excavating AI" text accompanying the exhibition they write: "ImageNet is an object lesson, if you will, in what happens when people are categorized as objects" (2019).

London's Photographers' Gallery has also thoroughly and provocatively explored the politics and ethics of image training sets in their year-long programme of events and commissions entitled "Data / Set / Match", led by curators

Katrina Sluis and Jon Uriarte and running over 2019-2020. The programme included exhibiting 14,197,122 photographs from ImageNet on the gallery's Media Wall, the images cycling through at a rate of ninety milliseconds per image following a computer script written by Malevé. *The Future Is Here!*, a video work by Mimi Onuoha commissioned by the gallery, explores the exploitation of labour involved in the annotation of training sets, a process often involving a dispersed group of crowdsourced workers connected through micro-tasking platforms such as Amazon's Mechanical Turk. Onuoha's video depicts the otherwise unseen domestic working spaces of these poorly compensated image taggers, many based in Venezuela. As Florian A. Schmidt describes in his response to Onuoha's video, "they work as freelance sub-sub-contractors, switching back and forth between different platforms that funnel the work from supranational corporations to people in the Global South" (2020). The artist Anna Ridler (who also featured in the "Data / Set /Match" programme) has likewise confronted the problematic ethics of training sets, both in terms of the labour practices involved in their creation and the classification systems they draw on. In works such as *Fall of the House of Usher* and *Mosaic Virus*, Ridler insists on producing her own datasets, employing machine learning systems trained on thousands of images she painstakingly creates herself. For the *Mosaic Virus* project, for example, Riddler photographed and hand classified over ten thousand tulips acquired during a single tulip season in Amsterdam.

Deep learning and the city

The critical attention focused on deep learning technologies by artists, designers and curators, in recent years in particular, has done much to expose the complex processes and infrastructures that underpin the purported AI revolution now underway. Understandably, many of these projects have placed the human at the centre of their investigations – questioning the systems of categorization, surveillance and deception machine learning may engender, as well as the precarious labour practices that enable their creation. And the role played by image training sets, the often-unseen foundations or 'ground truth' of deep learning, has justifiably attracted particular scrutiny. The ways in which computer vision and machine learning technologies are transforming urban space may have received comparably less attention from artists and designers, but the questions of automation, surveillance and classification that these projects address are of course also deeply connected to spatial concerns. As Shannon Mattern aptly states in her examination of the growth of intelligent mapping technologies, "with the stakes so high, we need to keep asking critical questions about how machines conceptualize and operationalize space. How do they render our world measurable, navigable, usable, conservable?" (2017). With this call in mind, the final section of this chapter will outline three recent art and design projects that do take as their primary focus the built environment and architectural or urban space, namely: the *Uncanny Rd.* online generative tool, Simone C. Niquille's CGI-based film *Homeschool*, and a trio of Forensic Architecture investigations: *Triple-Chaser*, *The Battle of Ilovaisk*, and *Model Zoo*.

Uncanny Rd. is a web tool designed by software developers Anastasis Germanidis and Cristóbal Valenzuela, the co-founders of RunwayML, a popular machine learning programme aimed at artists and designers. The project involves a relatively simple interface that provides users with a coloured map of a street scene which can be populated, according to preference, with a number of different object labels, such as streetlamps, pedestrians, cars, etc. This "semantic map" showing only the basic outline of objects within the scene is synthesized by a GAN trained on city streets, generating a somewhat distorted or impressionistic image of a streetscape with a slightly post-apocalyptic aesthetic – something reminiscent of *Mad Max* or the *Borderlands* videogame franchise. The project is described on the site itself as: "Collectively hallucinating a never-ending road using Generative Adversarial Neural Networks." Apart from being an amusing interactive drawing tool that showcases some of the generative capabilities of GANs, *Uncanny Rd.* is perhaps more significant for drawing attention to the training set it relies on, the Cityscapes Dataset.[3] Produced by the Max Plank Institute, TU Darmstadt, and Daimler AG R&D (the research arm of Mercedes-Benz), Cityscapes is an annotated or labelled dataset of recorded stereo video sequences captured in streets from fifty cities, mostly located in Germany. A Mercedes hood ornament appears at the bottom of every image produced by the *Uncanny Rd.* site, a giveaway as to the origins of the neural network's training material.

The motivation for producing Cityscapes, clearly not to enable the creation of playful online drawing tools, is made quite explicit in an accompanying research paper describing the dataset as "specifically tailored for autonomous driving in an urban environment" (Cordts et al. 2016, 1-2). To this end, Cityscapes provides "semantic urban scene understanding", or put more simply, it identifies and categorizes objects that appear in its large video collection of street scenes. The "semantic" object labels *Uncanny Rd.*

3 My thanks to Bernd Behr for sharing his insights on the significance of the Cityscapes Dataset.

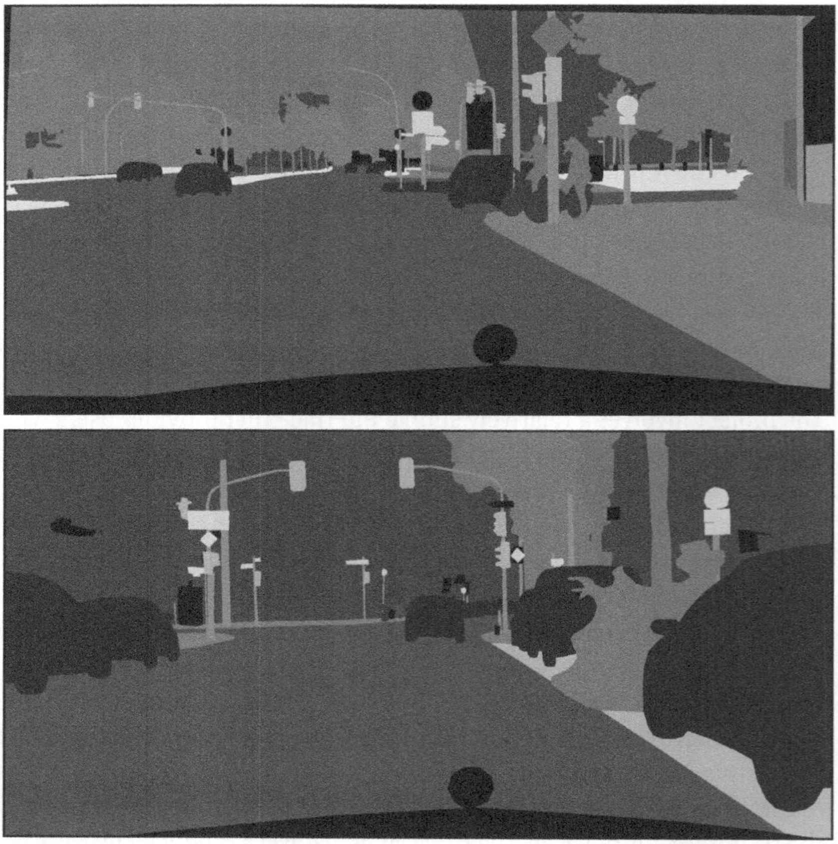

Figure 2.1. Semantic maps from the Cityscapes Dataset (Hamburg and Dusseldorf).

makes available to its users are pulled directly from the "class definitions" established in the Cityscapes Dataset, categories ranging from sidewalk, to bicycle, to person, to guard rail – all things that might be very useful for an autonomous vehicle to be able to recognize with a high degree of accuracy. The semantic mapping of Cityscapes is thus one component (along with detection technologies like Lidar) of the complex "sense-making capacities" of autonomous vehicles carefully considered by Sam Hind in his chapter in this volume. While many of the datasets used in the computer vision research of autonomous vehicle companies are proprietary, Cityscapes has had a wider general influence due to its public availability. It surfaces in a number of additional research areas, for example, in the video-to-video synthesis work conducted by Ting-Chun Wang and others at Nvidia and MIT, research that takes *Uncanny Rd.* a step further by generating photo-realistic moving video from the semantic maps that Cityscapes enables (Wang et al. 2018).

While perhaps not as obviously problematic as the issues of racial and gender bias inherent in a dataset like ImageNet, urban training sets such as Cityscapes nevertheless raise related questions of classification and standardization. Is it important, for example, that this widely influential dataset is based exclusively on scenes from German cities? What unintended consequences might arise from the public reliance on a Daimler AG produced training set, beyond the branding effect of the omnipresent Mercedes logo in every image generated from Cityscapes image data? What influence will a classification system attuned to the specific goals of autonomous vehicle design have on other forms of urban research making use of the dataset? Fiona McDermott articulates some of these concerns in her thoughtful work on the kinds of sensorial regimes produced by autonomous vehicle development. She writes that autonomous vehicles, "are only possible given huge amounts of collected and processed data, which begs the question as to how these exhaustive amounts of information might in turn have implications for the design and use of the space" (2019, 252). McDermott references the cautionary analysis of Florian Cramer who finds it all too easy to imagine an urban environment designed to be optimized for the limited category recognition of our current machine vision systems: "[A] ll cars and highways could be redesigned and rebuilt in such a way as to make them failure-proof for computer vision and autopilots. For example, by painting all cars in the same specific colors, and with computer-readable barcode identifiers on all four sides, designing their bodies within tightly predefined shaper parameters to eliminate the risk of confusion with other objects." (in McDermott 2019, 252).

What's clear from examples like Cityscapes Dataset is that computer vision technologies and the neural networks they rely upon are not only producing new machinic readings of the city, they are also altering the way humans view and interpret their urban surroundings. For Steve F. Anderson the current task is not to reinforce an opposition between organic human seeing and machine vision, given how inevitably intertwined the two have become, but instead to reflect on the ways human vision has been "reconstituted in dialogue with the computational" (2017, 82). However alien or uncanny the semantic maps or GAN-produced images of machine vision may appear, the forms of information they prioritize and the particular ways in which they segment and order the world shapes, for better or for worse, our own patterns of seeing and urban understanding. As the media philosopher Vilém Flusser noted of the computational images emerging in the 1970s and 80s, our technical images don't simply represent the outside world, they also envision or inform it: "Technical images are not mirrors but projectors" (2011, 51). The images used and produced by deep learning networks constitute some of the most important technical images of our current age and they undoubtedly project a specific regime of computational vision on the contemporary city.

The precarious networks of labour involved in other processes of image classification are also very much present within the computer vision research of the automotive industry. In fact, Schmidt's research on the human workers teaching self-driving cars "to see" reveals the emergence of a new sector of specialist platforms catering specifically to the labour demands of deep learning dependant industries like autonomous vehicles. He notes, "probably the most important lesson from studying the crowdsourced production of AI training data is that in the relatively short time of one and a half years the automotive industry was able to access hundreds of thousands of new workers, through a labour supply chain of venture capital funded platforms which sprung up like mushrooms to cater for this new demand" (Schmitt 2019, 25). This dispersed network of urban workers, predominantly from the global south, is a less frequently acknowledged geographic by-product of this developing technology.

The impact of deep learning and machine vision on design and automation is being played out on multiple urban scales, ranging from the metropolitan to the domestic. The recent work of designer Simone C. Niquille moves us from a concern with autonomous mobility in the city to a consideration of the technologies of automation targeting interior space. Her animated film *Homeschool* (2019) exposes yet another image dataset, this time one used in the computer vision training of domestic robots. The film is set within the CGI interior of a home populated with rendered objects derived from SceneNet RGB-D, a training set produced by the Dyson Robotics Lab at Imperial College. In this case the dataset is comprised of computer generated or "synthetic" images rather than photographs or videos, as this presents a more effective way of producing the mundane scenes of domestic clutter that an automated vacuum cleaner, for example, might rely on in order to learn how to navigate its environment. After all, we don't tend to offer up photographs of our messy living rooms on Flickr, or at least not in the vast quantities required for deep learning.

Niquille's film was originally titled *Regarding the Pain of Spotmini*, referencing the smaller iteration of the dog-like Spot robot produced by Boston Dynamics, this miniature version being small and nimble enough to handle the confined spaces of domestic and office interiors. Using a method that can appear a little surreal, SceneNet RGB-D produces its database of images by allowing synthetic objects to randomly drop from the ceiling of a CGI room,

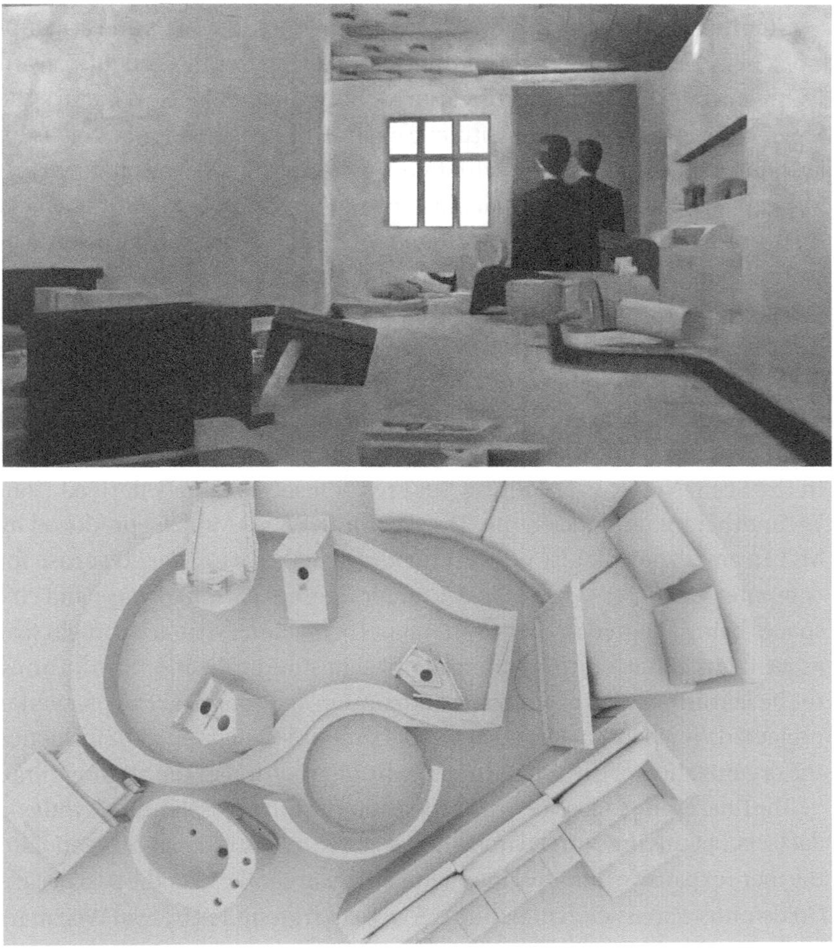

Figure 2.2. *Homeschool* (2019) by Simone C. Niquille. Courtesy of Simone C. Niquille.

settling according to the gravitational logic of a physics engine. Niquille's film presents the viewer with the anthropomorphized inner monologue of a robotic computer vision system as it "learns what a home is". The robotic protagonist moves about the space becoming gradually more proficient at naming objects like doors, plants and furniture. In a humorous, but also slightly sinister moment, the vision system approaches a CGI handgun lying on the floor of a living room that also contains a dining table and a child's pram. "Decoration? Toothbrush? Candle?" the voice asks, apparently struggling to identify the synthetic object. As the voice self-reflexively comments at the conclusion of the film: "The limits of my categories mean the limits of my world."

Niquille's interest lies in exploring the numerous decisions and assumptions of language that underpin something as apparently straightforward as the description and categorization of household objects. What are the logic parameters of what constitutes a chair in the eyes of a computer vision system? A piece of furniture with four legs? Anything we can sit down on? She explains, "Autonomous machines' computer-vision capabilities depend on the resolution of their training database. The database, however, is a subjective collection created by engineers, technicians or academic researchers. Once filtered through computer vision, this subjectivity becomes obscured: the seeing technology is too easily mistaken as an impartial agent" (Niquille 2019, 90). The inevitable tendency towards standardization involved in these systems is also an important consideration for Niquille. In a kind of recursive loop of uniformity, she reveals that the rendered objects included in the SceneNet RGB-D training set are themselves largely derived from yet another image dataset, the "Dataset for IKEA 3D Models" produced by MIT in 2013. The ubiquity of IKEA furniture makes it an ideal test case for computer vision research. Just as Cramer foresees cars, highways and city spaces being adapted to the requirements of machine vision, we might just as easily imagine a future of interior design standardization conforming to the learning needs of domestic automation and robotics. The particular projected viewpoint of neural networks thus has the potential to influence the organization of the urban from the infrastructural to the architectural.

The final example considered in this chapter also involves the use of synthetic datasets, but this time turned from the restrictive sphere of domestic interiors to the more expansive terrain of international urban conflict. For the past decade Goldsmiths' Forensic Architecture (FA) research group, led by Eyal Weizman, have employed advanced visualization technologies like digital animation and simulation in their important investigations of human rights violations, political violence and issues of environmental justice (Weizman 2017, McKim 2017). The incorporation of deep learning and computer vision techniques into the group's research methods is a more recent development, one supported by the arrival of FA members like software developer Lachlan Kermode.

The first demonstration of these new approaches can be seen in the agency's *Triple-Chaser* film, FA's response to an invitation to participate in the controversial 2019 Whitney Biennial. The exhibition had already been boycotted by a number of invited artists, a protest against the involvement of Whitney board vice-chairman Warren B. Kanders, whose company the Safariland Group produced tear gas munitions used by US agents against migrants at the US-Mexico border in an incident on November 25, 2018. The FA film, narrated by the musician David Byrne, documents the group's

Figure 2.3. *Triple-Chaser* (2019) by Forensic Architecture/Praxis Films. Courtesy of Forensic Architecture/Praxis Films.

process of training a machine learning classifier to search for images of the "triple-chaser" tear gas grenades manufactured by Defense Technologies, a subsidiary of Safariland. Able to locate only a hundred images of the triple-chaser grenade online (far too few to serve as a functional training set), FA turned to generating a synthetic image data set as a method of training their machine learning system.[4] Based on video footage of triple-chasers provided to FA by artists and activists and specifications available

4 A detailed account of the group's use of synthetic images is available in the FA report "Synthetic Data Generation: Development of Data Classification Tools".

in product catalogues, the group was able to create a digital 3D model of the grenade which could then be inserted into various background images (both computer generated and photo-realistic) in order to build a sizeable training set. Some of these images were produced using a process not unlike the one used to generate the images in the SceneNet RGB-D dataset, dropping CGI triple-chaser grenades randomly into scenes in order to produce a large variety of possible configurations. Having trained their machine learning system to identify the triple-chaser, FA is now deploying the classifier to search for the grenades across online images and video repositories, such as YouTube. The list of places where the group has already identified the use of Safariland-produced grenades against civilians is already long and includes Turkey, Peru, Iraq, Yemen, Egypt and Palestine, amongst other countries.

The *Triple-Chaser* film was both a provocation to the Whitney and an opportunity for FA to prototype a new method of research. A synthetic image approach to machine learning has since been employed in at least two subsequent investigations. *The Battle of Ilovaisk* investigation, commissioned by the European Human Rights Advocacy Centre (EHRAC) and the Ukrainian Legal Advisory Group (ULAG), called on FA to gather and analyse available evidence of the presence of the Russian military in Eastern Ukraine during a battle in the summer of 2014 between pro-Russian separatists and the Ukrainian Armed Forces. FA again experimented with the use of a machine learning classifier to help automate the process of analysing a large amount of open source information. This time the machine learning system was trained to recognize Russian military vehicles, such as the T-72B3 tank. Once trained, the classifier could then be programmed to automatically scour video platforms like YouTube.

Finally, FA's *Model Zoo* initiative, undertaken in collaboration with Bellingcat and Amnesty International, is the ongoing development of an open-source library of 3D models of weapons and munitions, along with various classifiers trained to identify them. A possible shared resource for multiple human rights organizations, the *Model Zoo* project confronts some of the barriers of access to deep learning technologies faced by non-commercial institutions. As will by now be clear, the effectiveness of machine learning in any domain is largely dependent on the availability of suitable training sets, which are expensive to produce and limited by image attainability. As a result, the production of datasets has been heavily weighted towards applications with the potential for large economic payoffs such as autonomous vehicles or industrial robotics. The *Model Zoo* initiative by FA is an attempt to ensure that the potential of deep learning technologies is not limited to either commercial ventures, with often problematic labour consequences, or even more troubling forms of control or surveillance. The group's forays into machine learning therefore echo FA's longer tradition of turning the advanced visualization technologies that are too often the exclusive domain of state powers and corporate interests towards a decidedly different agenda of human rights activism. While deep learning technologies are already shaping the built environment on multiple levels, Forensic Architecture's experiments introduce the potential for a productive machine vision intervention in urban conflict zones with substantial geo-political implications.

Conclusion

The projects outlined above provide at least an indication of how deep learning technologies are already impacting the design, organization and occupation of cities. These works provoke specifically urban or architectural questions, while also raising issues that are present across a wider field of art and design concerned with machine learning and AI. The critical projects of the past several years have done much to expose the inner working and inherent pitfalls of the training sets and computer vision systems employed in human oriented machine learning systems. In spatially oriented fields ranging from driverless vehicles to domestic robotics, we find equivalent problems of bias, classification, and automation. In her insightful book *Cloud Ethics* Louise Amoore asserts that the most pressing ethicopolitical questions arising from neural networks are less those related to the common fears of automation breaking free from human control and more those occasioned by "a machine learning that generates new limits and thresholds of what it means to be human" (2020, 65). The examples highlighted in this essay reframe this question slightly, compelling us to ask what it now means to be human in an urban environment increasingly shaped by machine vision.

Whether through detailing technical histories or producing creative investigations there remains work to be done to better comprehend and contend with technologies that are having an undeniably transformative impact on contemporary visual culture and urban life. The most promising of these projects are not only critiques, they are also efforts at greater understanding and explorations of alternative applications. Niquille's *Homeschool*, for example, literally gives voice to the machinic intelligences increasingly embedded within our domestic spaces, while the work of Forensic Architecture encourages us to challenge the current use of these emergent technologies by envisioning ways to deploy them towards different and unanticipated political ends.

References

Amoore, Louise. 2020. *Cloud Ethics: Algorithms and the Attributes of Ourselves and Others*. Durham: Duke University Press.

Anderson, Steve F. 2017. *Technologies of Vision: The War Between Data and Images*. Cambridge, Mass: MIT Press.

Bishop, Christopher M. 2006. *Pattern Recognition and Machine Learning*. New York: Springer.

Convolutional Neural Networks. NIPS Proceedings.

Cordts, M. Omran, S. Ramos, T. Rehfeld, M. Enzweiler, R. Benenson, U. Franke, S. Roth, and B. Schiele. 2016. "The Cityscapes Dataset for Semantic Urban Scene Understanding." *Proceedings of the IEEE Conference on Computer Vision and Pattern Recognition* (CVPR).

Crawford, Kate and Trevor Paglen. 2019. "Excavating AI: The Politics of Images in Machine Learning Training Sets." https://www.excavating.ai/

Flusser, Vilem. 2011. *Into the Universe of Technical Images*. Minneapolis: University of Minnesota Press.

Forensic Architecture. 2018. "Synthetic Data Generation: Development of Data Classification Tools." https://synthetic-datasets.sfo2.digitaloceanspaces.com/website/FA_Synthetic-Data-Generation_Report.pdf

Fukushima, Kunihiko. 1980. "Neocognitron: A Self-organizing Neural Network Model for a Mechanism of Pattern Recognition Unaffected by Shift in Position." *Biological Cybernetics* 36 (4): 193–202.

Gershgorn, Dave. 2017. "The Data that Transformed AI Research — And Possibly the World." *Quartz* https://qz.com/1034972/the-data-that-changed-the-direction-of-ai-research-and-possibly-the-world/

Goodfellow, Ian, Yoshua Bengio and Aaron Courville. 2016. *Deep Learning*. Cambridge, Mass: MIT Press.

Goodfellow, Ian, Jean Pouget-Abadie, Mehdi Mirza, Bing Xu, David Warde-Farley, Sherjil Ozair, Aaron Courville, Yoshua Bengio. 2014. "General Adversarial Nets." NIPS Proceedings.

Haraway, Donna. 1995. "Universal Donors in a Vampire Culture: It's All in the Family. Biological Kinship Categories in the Twentieth-Century United States." In *Uncommon Ground: Rethinking the Human Place in Nature*, edited by William Cronon, 321-377. New York: W. W. Norton.

Harvey, Adam. 2019. "MegaPixels: Face Recognition Training Datasets." https://ahprojects.com/megapixels/

Hubel, David H. and Torsten Wiesel. 1959. "Receptive Fields of Single Neurones in the Cat's Striate Cortex." *Journal of Physiology* 148 (3): 574–91.

Krizhevsky, Alex, Ilya Sutskever and Geoffrey Hinton. 2012. "ImageNet Classification with Deep

LeCun, Yann, Patrick Haffner, Léon Bottou and Yoshua Bengio. 1999. "Object Recognition with Gradient-Based Learning." In *Lecture Notes in Computer Science, Vol 1681*. Berlin: Springer.

Mackenzie, Adrian. 2017. *Machine Learners: Archaeology of a Data Practice*. Cambridge, Mass: MIT Press.

Mattern, Shannon. 2017. "Mapping's Intelligent Agents." *Places Journal*. https://placesjournal.org/article/mappings-intelligent-agents/

McCulloch, Warren S. and Walter Pitts. 1943. "A Logical Calculus of the Ideas Immanent in Nervous Activity." *Bulletin of Mathematical Biophysics* 5: 115-133.

McDermott, Fiona. 2019. "New Sensorial Vehicles: Navigating Critical Understandings of Autonomous Futures." In *Architecture and the Smart City*, edited by Sergio M. Figueiredo, Sukanya Krishnamurthy, Torsten Schroeder, 247-251. London: Routledge

Malevé, Nicholas. 2019. "An Introduction to Image Datasets." *Unthinking Photography*. https://unthinking.photography/articles/an-introduction-to-image-datasets

McCorduck, Pamela. 2004. *Machines Who Think*. Natick, Mass: A K Peters.

McKim, Joel. 2017. "Speculative Animation: Digital Projections of Urban Past and Future." *Animation* 12 (3): 287–305.

———. "Trevor Paglen Trains His Sights on the Rise of Machine Vision." *Apollo Magazine*. https://www.apollo-magazine.com/trevor-paglen-machine-vision/

Minsky, Marvin and Seymour A. Papert. 1969. *Perceptrons: An Introduction to Computational Geometry*. Cambridge, Mass: MIT Press.

Nielsen, Michael. 2019. *Neural Networks and Deep Learning*. http://neuralnetworksanddeeplearning.com/chap1.html.

Olazaran, Mikel. 1996. "A Sociological Study of the Official History of the Perceptrons Controversy." *Social Studies of Science* 26 (3): 611–659.

Niquille, Simone C. 2019. "Regarding the Pain of SpotMini: Or What a Robot's Struggle to Learn Reveals about the Built Environment." In *Machine Landscapes: Architectures of the Post-Anthropocene*, edited by Liam Young, 84-91. Oxford: Architectural Design.

Paez, Danny. 2019. "'This Person Does Not Exist' Creator Reveals His Site's Creepy Origin Story." *Inverse*. https://www.inverse.com/article/53414-this-person-does-not-exist-creator-interview

Schmidt, Florian A. 2019. "Crowdsourced Production of AI Training Data: How Human Workers Teach Self-Driving Cars to See." Dusseldorf: Hans-Böckler-Stiftung.

———. (2020). "Unevenly Distributed." *Unthinking Photography*. https://unthinking.photography/articles/unevenly-distributed

Wang, Ting-Chun, Ming-Yu Liu, Jun-Yan Zhu, Andrew Tao, Jan Kautz and Bryan Catanzaro. 2018. "High-Resolution Image Synthesis and Semantic Manipulation with Conditional GANs." *Proceedings of the IEEE Conference on Computer Vision and Pattern Recognition* (CVPR).

Wei, Jerry. 2019. "AlexNet: The Architecture that Challenged CNNs." *Medium*. https://towardsdatascience.com/alexnet-the-architecture-that-challenged-cnns-e406d5297951

Weizman, Eyal. 2017. *Forensic Architecture: Violence at the Threshold of Detectability*. New York: Zone Books.

About the author

Joel McKim is Senior Lecturer in Digital Media and Culture and the Director of the Vasari Research Centre for Art and Technology at Birkbeck, University of London. He is the author of *Architecture, Media, and Memory: Facing Complexity in Post-9/11 New York* (Bloomsbury 2018) and recently completed a visiting fellowship at the V&A Museum, working on the research project "A Pre-History of Machine Vision".

3. Machinic Sensemaking in the Streets: More-than-Lidar in Autonomous Vehicles

Sam Hind

Abstract

In recent years, lidar has increasingly been deployed in the testing of prototype autonomous vehicles. Rather than mapping forest cover or urban terrain, however, lidar has been used to map driving environments. This chapter explores the machinic sensemaking capacities of prototype autonomous vehicles, both composite as well as "distributed", with various, interconnected sensing systems and software programmes used for orientation, perception, and decision-making. In this, vehicles draw on sensing technologies with different observational ranges, prioritizing some over others at particular distances. Yet enabling this machinic sensibility involves undervalued, and misunderstood, visual responsibilities assumed by so-called "vehicle operators" during tests. Without this important work, prototype autonomous vehicles risk ignoring, or mis-sensing, other road users – with fatal consequences.

Keywords: sensing, machinic sensibility, recognition, distributed media

Introduction

Short for "light detection and ranging", lidar has historically been used for the aerial mapping of vegetation and for surveying urban environments and heritage sites. By emitting pulses of light that bounce back off surfaces and objects, spectral images called "point clouds" are generated, derived from millions of innocuous lidar pulses. In recent years, however, lidar has increasingly been deployed by car manufacturers and technology companies in the testing of prototype autonomous vehicles. Rather than mapping

Rose, G. (ed.), *Seeing the City Digitally. Processing Urban Space and Time.*
Taylor & Francis Group, 2022
DOI 10.5117/9789463727037_CH03

Figure 3.1. A stylised rendering of how lidar "sees", or senses, an urban environment. Courtesy of Velodyne Lidar.

forest cover, or urban terrain, lidar has been used as a principal sensing system to map driving environments, and aid the detection of other road users, signs, and lines.

In this chapter, I suggest that lidar is central to the "sovereign" (Bratton 2015; Gekker and Hind 2019; Pasquale 2017) sensemaking capacities of prototype autonomous vehicles, able to "configure territory and power" (Lovink and Rossiter 2019, 99) in new ways. When taken apart, the sensory capacities of prototype autonomous vehicles are both composite as well as "distributed", courtesy of various interconnected sensing systems and software programs used for three critical operations: orientation, perception and decision-making (McCosker and Wilken 2020). Lidar never acts alone; hence I use the phrase "more-than-lidar" to indicate that lidar is reliant upon an integrated suite of sensing systems.

It is often suggested that autonomous vehicles "see" (Davies 2018; Metz 2018; Stilgoe 2017), yet the way they see the world is manifestly different to other forms of (human and non-human) sight. Whilst greyscale point clouds generated by lidar show the world in a skeletal form, equally common technicolour renderings depict it as a kind of parallel hyperreality. Neither capture the urban environment as rendered in photographs, maps, or stylized illustrations (Figure 3.1). Instead, lidar and its ancillary sensing systems render the urban environment anew, in turn affecting how decisions are made within cities.

To address this newness, the chapter will build on Sun-ha Hong's (2016) concept of "machinic sensibility", to consider how autonomous vehicles "sense" rather than see. In this, I suggest that the autonomous vehicle entails four orders of sensing: from *feeling* the shape, texture and form of phenomena in the urban environment, through the rote *capture* of sense data, to the processual calculation of *meaning* from the processing of such data, before arriving at the execution of *good* or acceptable decisions.

The unceasing flow of information that characterize "distributed media" (Munster and Lovink 2005) is rarely the case with prototype autonomous vehicles. Whilst the distribution of machinic capacities can be seen to generate endless successful relays of integration, offering greater fidelity to the sensed environment, this same distribution equally renders relays of *disintegration*. Here, erroneous classifications and clashing system priorities render sensemaking an unevenly distributed activity.

Distinct and *distant* capacities are operationalized through this distribution. In this, distance – most notably, the distance between vehicle and object(s) – becomes a significant spatial principle through which judgements are made, and decisions executed. Yet, for the distant capacities of worldly phenomena to become useful, sensing units within such a distributed system must be *prioritized*, such that some assume greater significance at specific moments, or in specific situations. In this, the capacities of other road users, road surfaces, or entire junctions or road layouts are mobilized in ways that might otherwise not be, with these priorities encoded into the protocols of onboard software.

It is this uneven distribution of machinic capacities that is reflected in the differentiation of "sociotechnical agency" (Rose 2017, 779) at an operational level. As Gabrys and Pritchard contend, sensing practices "shift attention to formations and processes of *experience* across multiple entities" (2018, n.p., emphasis added). As such, this chapter explores how sensemaking in autonomous vehicles generates a differentiation in the distribution of experience, affecting some in qualitatively different ways to others.

As the chapter proceeds, I consider different aspects of the sensemaking capacities of prototype autonomous vehicles. I begin by focusing on the technical features, and operational limits, of specific lidar products used in developmental autonomous vehicles, considering how different models and their possible configurations affect these capacities. I then move on to consider a crash in Tempe, Arizona in March 2018, involving a prototype autonomous vehicle operated by Uber Advanced Technologies Group (ATG), that killed a woman called Elaine Herzberg. I contend that the crash, and the subsequent investigation, revealed the contingencies of classification, as Herzberg was variously re-classified as different objects (car, bike etc.) but never accurately as a pedestrian, in the moments before the crash.

In the final section, I consider how the nominal "supervisor" of the prototype vehicle at the time of the same crash, Rafaela Vasquez, was committed to performing an array of duties meant to enable or "fine-tune" the eventual sensemaking capacities of the autonomous vehicle. By studying the US National Transportation Safety Board (NTSB) report, I query the significance of her own visual sensibilities, and her repeated glances towards the central console of the vehicle. The central console was where her personal mobile phone was allegedly stored, but also where a tablet computer was similarly placed, on which Vasquez was committed to record system errors and driving infractions made by the vehicle in autonomous mode.

Machinic Sensibility

As Gabrys argues, "usually, some version of a cognizing human is at the centre of work on sensing", with sensing "tied to particular types of human embodiment, engagement, and experience" (2019, 724). Nevertheless, as Gabrys continues to suggest that "sensing practices", as she refers to them, extend beyond the human to an often-complex arrangement of "sensing entities and modes of experience" incorporating "computational sensors that monitor environmental pollution, to organisms that sense and bioaccumulate environmental toxins, and satellite that remotely sense aquifers" (2019, 724).

In this chapter, I want to focus on a particular constellation of sensing entities that together form a kind of "machinic sensibility" (Hong 2016), within the "driving-machine" (Hind 2019) itself. Machinic sensibility, in Hong's definition, describes "technical objects' own ability to sense the material world, and derive information through this process, in ways that are always entangled with, but ultimately distinct from human sensibility" (Hong 2016, 15). Here, media are only "*indirectly* correlated to human modes of experience," in which "the avenue of their impact on human experience and of their implications of humans within their operationality has shifted from a direct to an indirect modality" (Hansen 2015, 6, emphasis in original). In Hong's words, "such engineering *entirely bypasses, occurs prior to, and in sensory regions inaccessible by*, the human subject" (Hong 2016, 15, emphasis in original).

Machinic sensibility, then, is defined by an operational agency in which kinds, or modes, of sensing occur without direct correlation to, or impact on, human experience. Thus, whilst Gabrys (2019) extends the notion of sensing *practices* beyond the strictly human, to all manner of other possible technological and biological agents, both Hong (2016) and Hansen (2015) point towards a different kind of sensing *operation* largely occurring beyond or outside the human, in which to some degree, sensemaking is automated and/or autonomous (Andrejevic and Burdon 2014). Here the point is not that human awareness of, or access to, these sensemaking procedures is entirely impenetrable; but that these sensing processes are functionally distinct and independent from (human) awareness or access. In other words, they do not require direct human involvement to engage in sensemaking activities. This is what Hansen alludes to when he discusses the "veritable inauguration of new, *properly technical* domains of sensation" brought into being through the development of "machinic sensors *that possess sensory domains of their own*" (2015, 54, emphasis added).

I argue that this machinic sensibility is dependent upon four orders, or interpretations of sensing, expanding on Hong's own two-fold distinction. Firstly, this sensibility is a process of *feeling*, in which the likely forms of phenomena are sensed. For lidar, this feeling is enacted at the point of contact between individual pulses of light and objects within the urban environment. Only after the return of many more pulses do such objects start to come into view, with shapes, textures and contours rendered increasingly visible as a lidar unit scans the landscape. Secondly, this sensibility also invariably entails *meaning* making, in which phenomena are made sense of, or understood. Within autonomous vehicles, as I will discuss, this meaning making is distributed, even if lidar is responsible for the bulk of the sensing.

Beyond these two definitions that Hong identifies, I argue that the term machinic sensibility also denotes a process of capture (Agre 1994; Gekker and Hind 2019), in which the form (feeling) and comprehension (meaning) of phenomena are recorded, stored, and utilized in order to enhance the vehicle's ongoing perceptive capabilities. Lastly, this sensibility is meant to arrive at a good decision; that is, a normative outcome deemed "sensible", as it is encoded into decision-making software. This final interpretation posits that sensemaking is not a neutral pursuit, based only on the application of established scientific principles (for example, lidar and the speed of light), or computational limits (image processing times), but guided by expectations, and conventions, on the "social road" (Brown and Laurier 2017).

Automated, or autonomous, sensing operations can thus be said to "broker human accessibility" to the urban environment, with machinic sensibility constituting a different "domain" of sensibility, in which meaning is derived differently (Hansen 2015, 6). This access, I will contend later, is brokered through novel modes of machinic supervision within the autonomous vehicle, as human drivers become expected to monitor, and document, otherwise "autonomous" sensing operations. Expanding on how machinic meaning making is distinctive, Bunz suggests:

> Artificial Intelligence [AI] systems specialized for object recognition in images [...] identify objects depicted in an image in a very particular way: they record the pixel formations i.e., edges and textures of an image, and its shades and different regions of colour, to then calculate statistically the highest possibility [for] what those formations of edges might illustrate. (2019, 272)

In this characterization of AI image recognition processes, AI systems do not interpret images in the same way as humans. Rather than scanning an image for things that we think resemble familiar objects (a human face, a tree, a building), AI systems trained in object recognition instead consider the properties of these objects as they are composed in the image itself. In such systems, Bunz continues, "meaning is not understood but *calculated*" (2019, 272, emphasis added), with meaning derived instead from statistical confidence or likelihood that an object in an image is as it is according to its properties. Thus, that the calculation of such meaning occurs through a kind of *feeling* in which edges, textures and shades become critical sources of information.

It is this calculated form of feeling that guides lidar, with systems capable of measuring the reflectance of surfaces based on the "intensity" of lidar returns. However, lidar's ability to offer such insight is necessarily shaped by the technical limitations of the type or model of lidar device. Typical products used in prototype autonomous vehicles include Velodyne Lidar's Puck and HDL-64E models. The Puck, as the name suggests, is shaped like a hockey puck and has a 100m range, "best-in-class accuracy and calibrated intensity" as well as a "sensor-to-sensor interference mitigation feature" (Velodyne Lidar 2020a). It is commonly used by manufacturers to provide additional lidar sensing support along the side of the vehicle. The HDL-64E, on the other hand, is a "high definition real-time 3D lidar" with an enhanced 120m range, sixty-four channels, a 360° horizontal field-of-view, capable of generating "up to around 2.2 million points per second" (Velodyne Lidar 2020b). It is typically used to provide principal lidar capabilities on the roof of the vehicle (as illustrated in figures 3.1 and 3.3), and can usually be identified by the rotating casing that exposes the sensors whilst in operation.

As a Velodyne Lidar executive has contended, "the resulting point cloud of distance and intensity information is so dense that computer programs can identify objects such as street curbs and overhead wires at distances of over

Figure 3.2. A lidar point cloud with return "intensity" visualized in colour. Courtesy of Velodyne Lidar.

100m" (Schwarz 2010, 429). However this claim, of "around 2.2 million points per second" is "configuration dependent" (Velodyne Lidar 2018a, 2). It is this configurative dimension that is central to the sensing capacities of the lidar model in question, allowing it to adapt, or be adapted, to different situations.

The HDL-64E can operate in two modes: single return and dual return. Single return mode only offers a density of around 1.3 million points per second (a less pointy cloud), where the lidar pulse simply records the first thing it hits (i.e. a "single" return). Dual return mode provides the magical figure of 2.2 million points per second, recording multiple hits instead. The latter, therefore, provides an evidently richer account of the urban environment.

On dual return mode, the manufacturer notes that "different environmental conditions require a different priority of the type of distance point returns" (Velodyne Lidar 2018b, 15). For instance, the unit can prioritize the "strongest" distance points (the default). Or, if desired, the last distance point returned can be prioritized. As further suggested, "poor visibility conditions, such as fog and dust, benefit from collecting the distance return values based on the 'last return' scenario". This means that the "near field occluding atmosphere is ignored", i.e. the area containing fog or dust (Velodyne Lidar 2018b, 15). This is another example of where the sensing capacity of the lidar model is configuration dependent. In a last return scenario, these "near things" are deliberately ignored, constructing an image of the urban environment that deliberately discounts the real-world presence of some objects.

Thus, both the *distance* of data collected and the *intensity* of data collected are contingent upon the calibration of the unit itself, radically transforming the ability of the lidar model to feel the urban environment, capture data on the nature of these interactions, derive meaning from them, and ultimately to execute good, or acceptable decisions.

Distributing capacities

However, this machinic sensing is not performed in a singular location, nor executed by a singular entity. Instead, machinic sensibility is dependent on the distribution of sensemaking capacities throughout the vehicle itself. Here I contend that this sensemaking is, firstly, spatially distributed: sensing not only takes place in different locations but is also "oriented" differently towards a surrounding environment. But, secondly, sensemaking is also informationally distributed: sensor data is variously distributed to different parts of the vehicle in order to execute acceptable decisions. In this section I consider how these distributive capacities might be conceived.

As Munster and Lovink (2005) write, "new media are increasingly distributed media", requiring a "distributed aesthetics" that "must deal simultaneously with the dispersed and the situated, with asynchronous production and multi-user access to artifacts [...] on the one hand, and the highly individuated and dispensed allotment of information/media, on the other". Sensemaking in the autonomous vehicle is predicated not only on such a distributed aesthetics, of which the asynchronous production of, and multiuser access to, images is the norm, but also by a distribution of capacities through which images can be produced. Thus, the sensemaking capacities of autonomous vehicles are more than a kind of "distributed cognition" in which "machines [...] operate with an autonomy that underwrites our need to rely on them without understanding them" (Hansen 2009, 310). In other words, the "complex distributions of cognition beyond consciousness" are enabled, but also made complex, by distributed sensemaking (Hansen 2009, 310).

More accurately, sensemaking in the autonomous vehicle is dependent on what Munster and Lovink refer to as "loops of dispersal", in which there is "no singular or 'end use' of/for information but rather the endless relaying of media, practices and experience as successive dispersals" (2005). Whether intentional or not, Munster and Lovink valorise both successive and *successful* loops of dispersal, in which the so-called "endless relaying" of media results in an indeterminable volume of differentiated images. I argue here, however, that whilst distributed sensemaking might embody Munster and Lovink's endless, successful relays, these capacities are perhaps better understood in reference to musical composer William Basinski's *The Disintegration Loops*. A set of ambient productions completed as the 9/11 attacks were happening, the records were made when Basinski attempted to digitize a set of analogue tape loops. Rather than a flawless transfer of original compositions made by Basinski in the 1980s, a series of altogether more ghostly recordings were produced as the metal coating on the tape loops proceeded to blister and physically disintegrate (Richardson 2012).

Sensemaking in the autonomous vehicle is very much dependent on an endless relay of information between sensing units, systems, and other physical components such as brake modules and steering wheels. In other words, loops of dispersal. However, in many situations, these relays do not always work as intended. Instead, they are better characterized as *loops of disintegration* as sensor units are wrongly calibrated, sensor data is poorly captured, objects incorrectly identified, and decisions wrongly executed. Yet rather than bringing these relays to a halt, like Basinski's tapes they generate entirely new forms: new point clouds, new "clusters" of data points (Amoore 2018), new trajectories, and ultimately new decisions.

Yet whilst machinic sensibility is dependent upon a sometimes-disintegrative distribution of capacities throughout the autonomous vehicle, it also engenders a "functional" (Pasquale 2017) or "infrastructural" (Bratton 2015) auto-*nomic* sovereignty (Gekker and Hind 2019) enabled by the reliability, accuracy, and comprehensive qualities of lidar. As Velodyne Lidar contends, using lidar alongside cameras and radar, "allows better field of view and makes more accurate localization and free space detection possible" (Velodyne Lidar 2018c, 6). Moreover, in low light conditions, "lidar significantly fill[s] in the gaps created by the limitations of [...] other sensors" (2018c, 6). In this, lidar's sovereign status is derived from its ability to produce more useful, nominally accurate, data in a variety of situations. The framing of lidar as a sovereign actor is not to suggest it either acts alone, or even acts at every decidable moment. Instead, it is to suggest that as a sovereign actor, other sensing systems work with, for, and under it. Whilst figure 3.3 elides the distributed nature of sensemaking in a prototype autonomous vehicle, it nonetheless illustrates lidar's sovereign status, to which other modes of sensing are typically subordinated. Rather than being non-existent or invisible, as in figure 3.3, these other modes offer critical support for sovereign sensemaking.

The issue of sovereignty and autonomous vehicles has typically been couched in moral terms, most evidently through the "moral machine" project (Awad et al. 2018) and the "trolley problem" (Ganesh 2017), in which decisions around who to "save" and who to "kill" are rendered in utilitarian terms.

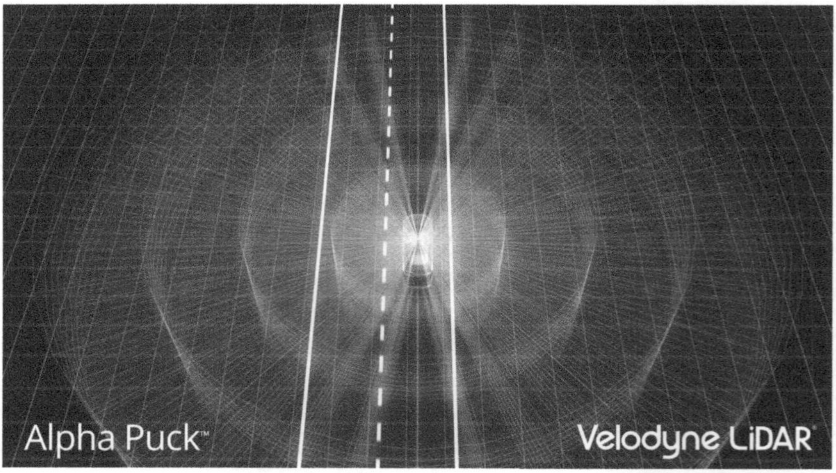

Figure 3.3. A stylised illustration of lidar's "sovereign" sensing capabilities, eliding its distributed nature. Courtesy of Velodyne Lidar.

Yet, limiting the discussion around machinic decision-making to moralistic debates ignores how the technical arrangement of sensing systems and attendant algorithmic software derive or calculate meaning, as discussed earlier. In this, there is no machinic desire to "make moral decisions" (Awad et al. 2018, 1); machines only desire arriving at acceptable decisions as they are calculated by onboard systems.

To consider how this distribution of capacities operates, I will turn for the first time to the Uber crash in Tempe, Arizona in March 2018. Here, I contend that the sovereign status of lidar is best explained in how sensor data captured of the urban environment is used to categorize other road users, as the bounding boxes in figure 3.1 show.

As Elaine Herzberg was walking across Northbound Mill Avenue in Tempe, Arizona, she was detected by an Uber ATG developmental automated driving system (ADS) onboard a modified Volvo XC90 test vehicle. To perceive the surrounding environment, the vehicle was equipped with 20 ultrasonic sensors, ten cameras, eight radar sensors, and one lidar unit (National Transportation Safety Board [NTSB] 2019a, 4). In the 5.6 seconds before Herzberg was hit, she was classified by the ADS on ten separate occasions, with each classification yielding a different possible trajectory Herzberg might take across the road (NTSB 2019a, 10-11).

On the first occasion, Herzberg was detected by the radar system as a *Vehicle*. 0.4 seconds later, she was detected by the lidar system and deemed to be a static object, putting her into the category of *Other*. One second later she is classified again as a *Vehicle*, but nonetheless is still presumed to be static. 2.6 seconds before impact, the ADS reclassifies her for a fourth time; this time as a *Bicycle*, deciding the bicycle by her side is being ridden. With 2.5 seconds left, the system finally predicts she is moving, yet through a lane adjacent to the test vehicle. 1.5 seconds before impact she is again classified as *Other*, and all previous trajectories are "reset". She is once again deemed to be a static object. At 1.2 seconds before impact, she is reclassified for a final time, now as a *Bicycle*, with the ADS predicting she is in the direct path of the test vehicle. Now too late to safely execute an emergency avoidance strategy, the ADS initiates "action suppression" designed merely to mitigate the effects of an impact. 0.2 seconds before Herzberg is hit, action suppression ends and the system issues an auditory warning. 0.02 seconds before impact, the vehicle operator (VO), Rafaela Vasquez, takes control of the steering wheel; now powerless to prevent the fatal crash (2019a, 10-11).

Here, sensemaking capacities are distributed variously. Firstly, through the processes of object detection and classification built into the ADS. With each subsequent classification – first as a *Vehicle*, then as *Other*, finally as a *Bicycle* – these capacities mutate, rendering Herzberg in different terms on each occasion. Secondly, between sensing systems in the vehicle itself, most notably between the radar system that first identifies Herzberg, and the lidar system that subsequently classifies, then reclassifies, her. In this, whilst the radar system is the first to pick Herzberg up, with its superior range detection, it is lidar that ultimately takes over as the vehicle approaches her. Thirdly, and belatedly, sensemaking capacities are distributed between the vehicle's sensing systems and the physical components designed to prevent a collision, such as the brakes or steering wheel. With this, the ADS communicates its decision, principally reliant upon the erroneous classifications based on lidar data, to the relevant components designed to perform the necessary actions. Then lastly, and even more belatedly, stepping outside of the intended, idealized, closed integration loop between these various sensing systems and physical components: the human VO herself contributes to the sensemaking capacities of the autonomous vehicle. Across these many capacities, sensemaking is not only distributed imperfectly, but catastrophically.

Here it becomes obvious that the vehicle in question did not, and was not, simply making a single moral decision at a nominal crossroads like in the fabled trolley problem. Instead, the system was engaged in an ongoing assessment of criteria, evaluating Herzberg at various stages, categorizing her differently each time, and making ongoing decisions to act (or not) on each occasion. At each stage, a different snapshot of the urban environment is made, with sensor data used to calculate the meaning of the objects in view. In sorting Herzberg into different categories the vehicle was reliant on the sovereign qualities of lidar. The tragic conclusion that can be drawn from this was that Herzberg was not moving "properly" or "normally" enough, or indeed, not moving "in the right place" within the urban environment, to be made sense of.

Distancing sense, prioritizing "recency"

Autonomous vehicles are being "computationally optimized for terrains [...] incorporating the sensing of elemental, atmospheric, and meteorological phenomena" (Hind 2019, 402). Consequently, as Gabrys and Pritchard (2018) argue, "distinct affective and political capacities are operationalized through [such] sensing practices". I want to argue here that not only are *distinct* capacities operationalized through the sensing operations of the autonomous vehicle, as articulated in the previous section. But in addition, that *distance* – most notably, the distance between vehicle and object(s) – becomes a significant variable in how these capacities are operationalized, as made evident in the death of Elaine Herzberg.

Ash argues, in reference to the Tesla Model S, that it is unhelpful to "understand smart objects' sensory capacities in the form of metrical distance" (2018, 170), despite it being used to promote the vehicle's "autopilot" driver-assist feature. Ash contends that such systems should be "defined by their capacity to *differentiate* between objects and *assign* the correct references to [...] objects to make distance sensible and intelligible" (2018, 170, emphases added). Metrical distance alone is no measure of the "smartness" of an object, nor indeed, of its sensemaking capacities. As Ash reiterates, "it does not matter how 'far' a sensor can reach, if that sensor cannot differentiate between objects [...] and so enable a car or driver to assign the correct references to those objects" (2018, 170).

To add to Ash's analysis, it is important to recognize that whilst the "smartness" of an object is not built (only) on its depth perception, neither is it based on universal perception. Autonomous vehicles are often touted as having "360 degree view" (Oxbotica 2019), or that specific systems can provide "360° [...] coverage" (NTSB 2019a, 4), or can "detect objects in a 360-degree area" (NTSB 2019a, 5), as illustrated in figure 3.3. In these statements, distance is mobilized differently, as a capacity of the vehicle to offer comprehensive depth perception. What these claims elide, however, is not only the composite nature of this apparently seamless and "universal" perception, but also the varying perceptive depth offered in 360 degrees. In other words, purported 360 degree vision is offered only through the integration of multiple units with specific sensing capacities, which in doing so, create an uneven depth to this purported capacity. Some sensing units may offer greater depth (radar) than others (ultrasonic sensors), whilst some may necessarily overlap (forward cameras and lidar) whilst rendering distance differently (compare radar and lidar).

Thus, it is only through a technical comprehension of distance that object-recognition, and therefore object differentiation, occurs. In the case of Uber, this is made possible through what it calls a "prioritization schema" that promotes "tracking by certain sensory systems over others" (NTSB 2019d, 12). Such a schema is "also dependent on the recency of an observation", where recency is defined as the "more recent detection of an object" (NTSB 2019d, 12). In other words, that some sensing systems, and some detection events, are prioritized over others at any one time. This is whilst lidar units, such as the Velodyne Lidar Puck or HDL-64E models discussed before, are also calibrated to prioritize either the strongest or last distance point recorded. An acknowledgement of the contestability of such a schema was made by Uber, post-crash, when it announced it would change the way the system "fuses sensor information" when predicting object trajectories (NTSB 2019d, 13). In any case, both distance point prioritization and sensor system prioritization are critical features of the prototype autonomous vehicle.

Take, once again, the moments before Herzberg was hit. 5.6 seconds before impact, she is first detected by the vehicle's radar system. Two radar units provide forward scanning and can operate in two modes. Mode one, a long-range scan, has "an observational range of up to 180 meters with a 20-degree field of view", whilst mode two, a medium-range scan, has "an observational range of up to 65 meters with a 90-degree field of view" (NTSB 2019a, 5). As the report continues, the "radar processing units conduct the initial processing of the [sensed] data, which the ADS then uses to build and continually update the representation of the surrounding environment" (2019a, 5). Whilst it is unclear which mode was active at the time, Herzberg was recognized as a vehicle. Thus, at 5.6 seconds before impact, Herzberg's distant capacities are deemed to resemble a vehicle; likely because she is simply present in a vehicle lane. Nevertheless, mere (metrical) distance is enough for such a recognition to occur; distant capacities are operationalized through the sensing operations of the vehicle. Metrical distance matters because, computationally and operationally, the radar unit attached to the vehicle has a sensory limit; either up to 180 metres, or 65 metres, depending on the operative mode.

Yet in this integrated process, as contended, some sensor systems take priority. At the time of the crash, only the Uber ADS was active. However, the Volvo XC90 was also equipped with a parallel advanced driver assistance system (ADAS) called City Safety. Although not a fully automated driving system, City Safety is designed to detect pedestrians in urban environments; comprised of what Volvo calls Forward Collision Warning and Automatic Emergency Braking. When the vehicle was being used in manual mode, controlled by a VO, "all the Volvo ADAS components were active and operated as designed" (NTSB 2019a, 13). Yet when the Uber ADS was activated, "all Volvo ADS components were automatically disengaged" (NTSB 2019a, 13). Only the vehicle's passive safety technologies, such as seatbelt pretensioners and airbag deployment systems, "remained active" in autonomous mode (NTSB 2019a, 14).

Two reasons are given for why the Volvo system was deactivated at the time of the crash. Firstly, that because the Uber ADS and Volvo ADAS both used radar, there was a "high likelihood of misinterpretation of signals" (NTSB 2019a, 14) between both. Secondly, that in receiving braking commands from either system, the "vehicle's brake module [would] not [have] been designed to assign priority" to either system (NTSB 2019a, 14). Subsequently, two sets of unresolvable conflicts occur.

Firstly, there is an identified or presumed conflict between sensing approaches. Here the issue is not that each individual system uses different sensing methods (one using lidar, the other radar, for instance), but that both use the same approach, i.e. radar. Likely due to respective system configurations, radar data will be processed and made sense of differently by each system. The result is differently interpreted data of the same phenomena using the same method. Secondly, there is a conflict between composite automation/assist systems. Here the issue is that each individual system – Uber's ADS and Volvo's ADAS – will likely send similar commands to the various modules in the vehicle assigned to move physical components such as the brakes. The result is possibly conflicting commands issued to components not programmed to decide which to listen to or ignore.

Ultimately, this means some sensing units, and some composite systems, as well as some detection events, are prioritized over others. The consequence of these conflicts – presumed or actively identified – is that some modes of distancing are prioritized over others; meaning only some distant capacities are operationalized at any one time. Why this matters is that the capacities of other road users in the urban environment are only realized through some sensing systems, and those identified more recently assume greater priority. Understanding when and where particular modes are themselves prioritized is critical to articulating the effect of these sensing systems on how the urban environment is variously perceived at any one time, according to the registered, and classified, capacities of other road users.

Enabling machinic sensibility, or "what's in a glance?"

The result of both a distribution of capacities and a prioritization of sensing is a differentiation in experiential effects. In arguing that machinic modes of sensing constitute a different "domain" of sensing (Hansen 2015, 6), I have not intended to erase the involvement of human actors in the operation-at-large. Instead, as outlined before, I argue that these sensing operations "broker human accessibility" (Hansen 2015, 6) to the urban environment. As the first section of this chapter hinted at, human actors in such arrangements become supervisors, overseeing how the machine operates. This was a role performed by Rafaela Vasquez in the fatal Uber crash in Tempe, Arizona, but also by many other VOs employed by the company as nominal machinic supervisors. In this final section I want to draw attention to the specific experiences of Rafaela Vasquez as affected by the distribution of capacities at the time of the crash: both subject to, and an unwitting enabler of, machinic sensibility. In other words, the sensing operations of the Volvo XC90, equipped as it was with an in-development Uber ADS, were only made possible through the interventions, interpretations, and interactions of human operators like Vasquez – or, indeed, the lack thereof.

Firstly, as a VO, Vasquez was responsible for carrying out a range of tasks before, during, and after testing. When the vehicle was in autonomous mode, she would have been expected to do three things: (a) continuously monitor the state of the vehicle and the road (b) take control of the vehicle should a dangerous situation arise, and (c) document performance-related incidents. In order to train VOs to perform these tasks correctly, they are subject to a three-week "onboarding process" in multiple locations, where they are taught vehicle handling skills, and introduced to various scenarios to "test [...] [their] decision making skills and ability to interact with the vehicle controls" (NTSB 2019b, 3). Then, VOs are tested on company procedures and processes, before being "re-localized" in relation to state driving laws in Arizona, and introduced to Uber ATG's infraction policies and test routes. Although Vasquez completed the training in a slightly different order, she followed the same three-week training course, intended to equip her with the skills to be a VO.

Yet Vasquez was originally trained on passenger operations (as opposed to test operation) according to a pilot/co-pilot model. In this format, two VOs would be present in any one test vehicle. One VO would occupy the driver's seat, ready to take control if a situation arose. The other VO would occupy the front passenger seat, supervising the vehicle's path, whilst tagging and annotating issues on a laptop that might arise whilst the vehicle was in autonomous mode. In this configuration, the three principle tasks for each VO, as outlined above, would have been divided between two VOs: VO1 (pilot) principally responsible for (a) and (b), whilst VO2 (co-pilot) principally responsible for (c). However, in October 2017, things changed. As the report details:

> Uber ATG integrated much of the co-pilot's functions into the 'front seat control application' (FSCA) software, housed on a centre-dash mounted tablet computer in the SDV [self-driving vehicle]. The FSCA interface was the primary means for the VO to interface with the SDS [self-driving system]. Complex functions on the FSCA were locked out once the SDV was in motion, and according to Uber ATG, functions that were available to the VO while the vehicle was in motion only required one to two taps to complete. (NTSB 2019b, 3)

In short, Uber consolidated the role of pilot and co-pilot into one VO and the aforementioned FSCA software. The result was that tasks (a), (b) and (c) – continuous monitoring, possible control, and performance documentation – were now expected to be performed by a single VO, sitting in the driver's seat. Not long after, Vasquez was trained on the interface, beginning work as a single VO a month later. The previously distinct training paths of passenger operations and test operations were now combined to reflect these changes.

Thus, Vasquez and all other VOs were responsible for interacting with FSCA software on tablet, affixed to the centre dashboard of the vehicle. Moreover, VOs were still expected to complete interactive tasks while the vehicle was in motion. Whilst, as the excerpt above mentions, "complex functions" were "locked out" whilst on the move, VOs were still required to perform other functions requiring "one to two taps to complete" (NTSB 2019b, 3). The report details four such input types, including "tagging an object of interest", "notifying the engineering team of an on-vehicle issue", "tagging incidents or infractions" and "tagging when the SDS performs incorrectly" (NTSB 2019b, 8). Thus, whilst each function might only have required one or two taps, the combined occurrence of these problems could demand repeated interactions with the tablet.

These functions were visually represented on the interface itself. If a VO wanted to tag an object of interest, they could locate the "label" icon in the bottom-left corner of the screen. If there was an on-vehicle issue, the VO could tap the "ticket" icon at the bottom-centre of the screen. If the vehicle had been involved in an incident or infraction, the VO could tap the "attn" (attention) icon, again, alongside the ticket option. If the autonomous system had acted strangely (although not necessarily dangerously), then the VO could press the "autonomy" icon at the bottom-right corner of the screen. Thus, in order for the VO to perform their ordinary duties – namely, the documentation of vehicle performance – they would have to get used to tapping the dashboard-mounted interface whenever necessary. All logged incidents would then be dealt with by relevant ATG teams, responsible for fixing or updating the responsible features. Test iterations – and, specifically, the documentation of incidents during them – were critical stages in the development of the sensemaking capacities of the Uber autonomous vehicle. Without the recording of these incidents – possibly unencountered in other test modes or simulated situations – the vehicle system might well be worse at making decisions, recognizing other road users, or obeying local traffic laws.

For the VOs like Vasquez, attention would naturally be divided between road and interface, windscreen and dashboard. In the final report published after an eighteen-month investigation, the probable cause was given as "the failure of the vehicle operator to monitor the driving environment and the operation of the automated driving system because she was visually distracted throughout the trip by her personal cell phone" (NTSB 2019d, 59). In records obtained from video streaming providers (including Hulu), NTSB determined Vasquez "was continually streaming a television show between 9.16pm and 9.59pm [...] That period covered the entire crash trip, which included 39 minutes on a public road" (NTSB 2019d, 24). These conclusions were drawn despite Vasquez stating she had "placed her personal phone in her purse before driving, and that her company phone was on the passenger seat at the time of the crash" (NTSB 2019d, 24).

Here, the intention is to not disagree with the conclusions drawn by the NTSB about the crash, after which Vasquez was charged with negligent homicide (Levin 2020). Nor is it to believe Vasquez's account of the crash; that her personal phone was in her bag, placed on the back seat of the vehicle, both out of sight and out of reach. Rather, the intention is to make sense of the tasks required to be performed by any VO whilst the vehicle is in autonomous mode, and those not permitted, i.e. like using a personal mobile phone. In other words, this chapter seeks to identify the precise role of – and the specific risks taken by – a VO ultimately responsible for enabling the eventual sensemaking capacities of the autonomous vehicle.

As an interview with Vasquez suggests, the latest VO training "indicated that she [VOs] may look at the iPad for 5 seconds and spend 3 seconds tagging and labelling" (NTSB 2019c, 6). VOs were expected to look forward at all times, including (indeed, especially) when the vehicle was in autonomous mode. Yet, they were also expected to perform tagging and labelling tasks as regularly as required, with up to 8 seconds spent looking at, and interacting with, the central dash-mounted tablet. As interior photos show, the lower console area "where a cell phone could be placed" (NTSB 2019b, 7) was directly underneath where the tablet was mounted. The NTSB deduced:

> From the time the VO exited the parking lot to the time of the crash, the VO frequently glanced down towards the lower centre console area. The Tempe Police tabulated the number of glances the VO made towards the lower centre console area during a 27-minute window, from 9.31pm to 9.58pm. During this timeframe, the VO glanced down at the same spot 204 times, of which 166 instances were when the vehicle was in motion. The[y also] estimated that [...] the VO's eyes were averted from the roadway [for] approximately 32% of the time. (2019b, 7).

Much meaning is attributed to the "glances" made by Vasquez towards the lower console area, and the frequency at which these glances occurred during the time the vehicle was in autonomous mode. Yet glancing towards this area was not against Uber policy. Indeed, as has been suggested, it was part of the assumed role of any VO – to look towards, and interact with, a tablet mounted on the central dashboard whenever an incident arose that required documenting. Necessarily, in doing so, VOs would have to look away from the road ahead, and down towards the interior of the vehicle; as well as concentrating on making an accurate record of any encountered incident.

Thus, this shift in attention was part of Vasquez's – and any VOs – assumed responsibilities. Without taking such action – repeated glances, diverted attention, concentration, tapping, and tagging – the developmental Uber vehicle would be without critical operational insights derived from test situations. In other words, the vehicle would likely fall short – just like it did in this crash – of correctly sensing other road users, and adapting to their presence. The future sensemaking capacities of the autonomous vehicle being tested were dependent on routine glances, just not the kind Vasquez was deduced to have made.

Conclusion

In this chapter, I have argued that whilst lidar is central to the sensemaking capacities of prototype autonomous vehicles, this sensemaking is only made possible through the distribution of responsibilities throughout any such vehicle. Further, I have contended here that this sensemaking is only enabled through the involvement of human operators involved also in the correction, and verification, of machine-readable driving worlds. This capacity is what Hong (2016) refers to as "machinic sensibility", a process through which technical objects recognize things in the world, and derive information from this recognition. Importantly, machinic sensibility is entangled with other forms of human sensing, visual and otherwise – whether in the form of quality control, oversight, or decision-making. Nevertheless, this machinic sensibility is better characterized through the figure of the sensing operation held at arms-length from human intervention.

In this, I have suggested that the machinic sensibility of lidar in the prototype autonomous vehicle is dependent upon four orders of sensing. Firstly, through a process of *feeling* or the interpretation of the shape and form of phenomena. Secondly, and necessarily, through a process of recording and *capturing* such phenomena, so that this feeling can be made operational. Thirdly, enabling the processual making of *meaning* through which phenomena are "made sense of". Then, lastly, through the execution of *good* decisions – a normatively-derived outcome deemed "sensible" and reasonable to at least some of the involved parties.

Yet, the machinic sensibility of lidar in the prototype autonomous vehicle is not being singularly, and solely, performed by and in the lidar unit itself. Instead, this machinic sensibility is dependent on the distribution of sensemaking capacities throughout the vehicle. This, I have argued, involves both a *spatial* distribution between components capable of aiding the four orders of sensing (feeling, capturing, meaning, good) and an *informational* distribution in which data is variously distributed to enable the smooth execution of decisions. Sensing is distributed to verify and authenticate sovereignty, exemplifying a case of functional or infrastructural auto-nomic sovereignty.

This machinic sensibility, however, is also dependent upon the operationalization of distant capacities. In this, the nominal distance between any lidar-equipped vehicle and objects within the urban environment is a critical factor in their being sensed. This operationalization is referred to as a "prioritization schema" (NTSB 2019d, 12) in which objects closer to the vehicle are prioritized over those further away. Moreover, "recency" (NTSB

2019, 12) – or the more recent detection of an object – is given priority over objects sensed longer ago.

The distribution, and distance, of machinic sensibility is, I argue, dependent on its enabling. Here, under specific test conditions, machinic sensibility as an operation is surfaced, or made available to human operators. In such instances, these human operators – and the tasks they are required to perform – are not only actively shaped by the operational capacities of lidar, but also the various interfaces that allow them to interrogate these capacities during test situations. As such, I contend that this surfacing, or availability, structures and scripts the experience of those made responsible for fine-tuning the sovereign sensemaking capacities of autonomous vehicles.

Throughout this chapter I have drawn on both off-the-shelf lidar products, as well as the specific testing of developmental autonomous vehicle systems. Most notably, I have focused on the crash in March 2018 in Tempe, Arizona, involving a prototype autonomous vehicle, that killed Elaine Herzberg. In the first instance I have suggested that Herzberg was subject to the ongoing assessment of operational criteria that led to her being classified, and re-classified, as various objects – from a car to a bike – in the seconds before impact. In the second instance, I have argued that this ongoing assessment was dependent upon her own "distant capacities", being variously sensed by lidar and other perceptive systems in the prototype autonomous vehicle, at different times. In this, Herzberg was interpreted, captured, made sense of, and ultimately decided on differently, at different distances to the vehicle itself. Then, thirdly, I moved on to Rafaela Vasquez, the nominal operator of the prototype vehicle involved in the crash itself. Here, I contended that her role as a diagnostician of the sensemaking capacities of the vehicle led to scrutiny of the application of her tasks as a certified vehicle operator. In this, I have queried the significance of the "glance": the repeated actions Vasquez is alleged to have made that impaired her ability to take control of the vehicle in the seconds before the crash. The sensemaking capacities of these prototype autonomous vehicles are dependent upon the interpretive, and interactive, work of vehicle operators such as Rafaela Vasquez.

What I have sought to do in this chapter is to give colour to the sensing operations performed by a prototype autonomous vehicle, particularly to how it perceives urban space, and to highlight bundled processes and practices that coalesce around these operations. What is critical to note, therefore, is that the chapter has not speculated on any eventual or hypothetical sensemaking capacities of an autonomous vehicle. Instead, it has sought to articulate the sensing operations of prototype autonomous vehicles being tested at this moment, to make sense of how these operations are not only being performed, but also necessarily upgraded and improved. As such, it is a snapshot of the sensemaking capacities of "more-than-lidar" and the various loops of interpretation, meaning-making, and decision-making that comprise this arrangement. The perception of urban space – including perceiving it visually – is enabled or indeed disabled through these loops, in which particular objects are sensed, and made sense of, at any one time. When these loops short-circuit or disintegrate, as was the case in the Uber crash, sensemaking does not stop. Instead, novel, unintended, and potentially catastrophic effects result, generating a differentiation of experience, whether for other road users or those responsible for supervising the work of machines.

References

Agre, Phil. 1994. "Surveillance and Capture: Two Models of Privacy." *The Information Society* 10 (2): 101–27.

Amoore, Louise. 2018. "Cloud Geographies: Computing, Data, Sovereignty." *Progress in Human Geography* 42 (1): 4–24.

Andrejevic, Mark, and Mark Burdon. 2014. "Defining the Sensor Society." *Television & New Media* 16 (1): 19–36.

Ash, James. 2018. *Phase Media: Space, Time and the Politics of Smart Objects*. London: Bloomsbury.

Awad, Edmond, Sohan Dsouza, Richard Kim, Jonathan Schulz, Joseph Henrich, Azim Shariff, Jean-François Bonnefon, and Iyad Rahwan. 2018. "The Moral Machine Experiment." *Nature* 563: 59–64.

Bratton, Benjamin. 2015. *The Stack: On Software and Sovereignty*. Cambridge, MA: MIT Press.

Brown, Barry, and Eric Laurier. 2017. "The Trouble With Autopilots: Assisted and Autonomous Driving on the Social Road." *CHI '17: Proceedings of the 2017 CHI Conference on Human Factors in Computing Systems*, May: 416–29.

Bunz, Mercedes. 2019. "The Calculation Of Meaning: On the Misunderstanding Of New Artificial Intelligence As Culture." *Culture, Theory and Critique* 60 (3–4): 264–78.

Davies, Alex. 2018. "How Do Self-Driving Cars See? (And How Do They See Me?)." *Wired*, August 11, 2018.

Gabrys, Jennifer. 2019. "Sensors And Sensing Practices: Reworking Experience Across Entities, Environments, And Technologies." *Science, Technology, & Human Values* 44 (5): 723–36.

Gabrys, Jennifer, and Helen Pritchard. 2018. "Sensing Practices." In *Posthuman Glossary*, edited by Rosi Braidotti, and Maria Hlavajova, 394–96. London: Bloomsbury.

Ganesh, Maya Indira. 2017. "Entanglement: Machine Learning and Human Ethics In Driver-less Car Crashes." *A Peer–Reviewed Journal About* (*APRJA*) 6 (1): 76–87.

Gekker, Alex and Sam Hind. 2019. "Infrastructural Surveillance." *New Media & Society* 22 (8): 1414–36.

Hansen, Mark. 2009. "Living (with) Technical Time: From Media Surrogacy to Distributed Cognition." *Theory, Culture & Society* 26 (2–3): 294–315.

Hansen, Mark. 2015. *Feed-Forward: On the Future of Twenty-First-Century Media*. Chicago: The University of Chicago Press.

Hind, Sam. 2019. "Digital Navigation and the Driving-Machine: Supervision, Calculation, Optimization, And Recognition." *Mobilities* 14 (4): 401–17.

Hong, Sun–ha. 2016. "Data's Intimacy: Machinic sensibility and the Quantified Self." *Communication +1* 5 (3): 1–36.

Levin, Sam. 2020. "Safety Driver Charged In 2018 Incident Where Self-Driving Uber Car Killed a Woman." *The Guardian*, September 16, 2020.

Lovink, Geert, and Ned Rossiter. 2019. "Propositions on the Organizational Form." In *Organize*, edited by Timon Beyes, Lisa Conrad, and Reinhold Martin, 89–101. Lüneberg: Meson Press.

McCosker, Anthony, and Rowan Wilken. 2020. *Automating Vision: The Social Impact of the New Camera Consciousness*. New York: Routledge.

Metz, Cade. 2018. "How Driverless Cars See the World Around Them." *The New York Times*, March 19, 2018. https://www.nytimes.com/2018/03/19/technology/how-driverless-cars-work.html/.

Munster, Anna, and Geert Lovink. 2005. "Theses On Distributed Aesthetics. Or, What A Network Is Not." *The Fibreculture Journal* 7. http://seven.fibreculturejournal.org/fcj-040-theses-on-distributed-aesthetics-or-what-a-network-is-not/.

National Transportation Safety Board (NTSB). 2019a. "Highway investigation HWY18MH010: Vehicle Automation Report." *NTSB*, May 11, 2019. https://data.ntsb.gov/Docket/Document/docBLOB?ID=40477717&FileExtension=.PDF&FileName=Vehicle%20Automation%20Report-Master.PDF.

National Transportation Safety Board (NTSB). 2019b. "Highway investigation HWY18MH010: Human Performance Factual Report." *NTSB*, May 11, 2019. https://data.ntsb.gov/Docket/Document/docBLOB?ID=40477743&FileExtension=.PDF&FileName=Human%20Performance%20Factual-Master.PDF.

National Transportation Safety Board (NTSB). 2019c. "Highway investigation HWY18MH010: Uber ATG Vehicle Operator Interview." *NTSB*, May 11, 2019. https://data.ntsb.gov/Docket/Document/docBLOB?ID=40477745&FileExtension=.PDF&FileName=Human%20Performance%20Attachment%20-%20Uber%20ATG%20Vehicle%20Operator%20Interview-Master.PDF.

National Transportation Safety Board (NTSB). 2019d. "Highway Investigation HWY18MH010: Collision Between Vehicle Controlled By Developmental Driving System and Pedestrian, Tempe, Arizona, March 18, 2018." *NTSB*, May 11, 2019. https://data.ntsb.gov/Docket/Document/docBLOB?ID=40479021&FileExtension=.PDF&FileName=NTSB%20-%20Adopted%20Board%20Report%20HAR-19%2F03-Master.PDF.

Oxbotica (@Oxbotica). 2019. "Our Eyes Are A Vital Asset When It Comes To Spotting And Reacting To Danger On The Roads." *Twitter*, November 8, 2019. https://twitter.com/oxbotica/status/1192815056926457858.

Pasquale, Frank. 2017. "From Territorial to Functional Sovereignty." *Law and Political Economy Blog*, June 12, 2017. https://lpeblog.org/2017/12/06/from-territorial-to-functional-sovereignty-the-case-of-amazon/

Richardson, Mark. 2012. "The Disintegration Loops." *Pitchfork*, November 19, 2019. https://pitchfork.com/reviews/albums/17064-the-disintegration-loops/.

Rose, Gillian. 2017. "Posthuman Agency In The Digitally Mediated City: Exteriorization, Individuation, Reinvention." *Annals of the American Association of Geographers* 107 (4): 779–93.

Schwarz, Brent. 2010. "Mapping the World In 3D." *Nature Photon* 4: 429–30.

Stilgoe, Jack. 2017. "Seeing like a Tesla: How Can We Anticipate Self-Driving Worlds?" *Glocalism: Journal of Culture, Politics and Innovation* 3: 1–20.

Velodyne Lidar. 2018a. "HDL-64E Data Sheet." HDL-64E. https://velodynelidar.com/products/hdl-64e/.

Velodyne Lidar. 2018b. "HDL-64E User's Manual and Programming Guide." HDL-64E. https://velodynelidar.com/products/hdl-64e/.

Velodyne Lidar. 2018c. "A Safety-First Approach to Developing and Marketing Driver Assistance Technology." *Velodyne Lidar*, December 1, 2018. https://velodynelidar.com/wp-content/uploads/2019/08/ADAS-Whitepaperver.12.2018.pdf.

Velodyne Lidar. 2020a. "Puck: Compact, Powerful, Intelligence." https://velodynelidar.com/products/puck/.

Velodyne Lidar. 2020b. "HDL-64E: High Definition Real-Time 3D Lidar." https://velodynelidar.com/products/hdl-64e/.

About the author

Sam Hind is Research Associate in SFB1187 Media of Cooperation at the University of Siegen. He is co-editor of *Time for Mapping: Cartographic Temporalities* (Manchester University Press, 2018) and co-author of *Playful Mapping in the Digital Age* (Institute for Network Cultures, 2016). He has published in *Political Geography, Mobilities*, and *New Media & Society*.

4. Curating #AanaJaana [#ComingGoing]: Gendered Digital Lives and Networked Violence in Delhi's Urban Margins

Ayona Datta

Abstract

This chapter presents a gendered perspective on Delhi's urban future, produced and curated by young women living in its urban peripheries. Using the metaphor of #AnnaJaana [#ComingGoing] as a paradigm for the urban periphery, the chapter argues that the women's everyday mobility across the home, phone and the city highlight the ambiguities of their lives. Using WhatsApp diary entries of multimedia content (audio recordings, photographs, videos and text messages by women), conversations between the women and researchers as well as observations of the dynamics within the WhatsApp group over a period of six months, the chapter suggests that #AanaJaana highlights the ambiguities of living between digital-territorial exclusions and offers ways to speak back to the city from the margins.

Keywords: mobile phones, urban margins, gender, Delhi

Introduction

On January 1 2019, a month-long public exhibition titled #AanaJaana opened in Mandi House, the largest metro station in New Delhi. The exhibition presented a genre of co-curation by researchers, a graphic artist and young women living in Delhi's slum resettlement colonies. Curated almost exclusively from WhatsApp diary entries made by women from their mobile phones, #AanaJaana represented ways of belonging, identity, leisure, and the pervasiveness of gendered violence across digital, physical and social spaces in everyday lives of these women. As a creative practice that was

Rose, G. (ed.), *Seeing the City Digitally. Processing Urban Space and Time.*
Taylor & Francis Group, 2022
DOI 10.5117/9789463727037_CH04

Figure 4.1. Exhibition being installed in Mandi House. Author's own.

simultaneously intimate, embodied and collective, the exhibition once installed also turned the gaze of the women onto themselves, reflecting upon those networks of violence that have kept them excluded from urban development and infrastructural planning alike. #AanaJaana demonstrated the potential of a new visual language co-produced with the women, and its capacity to reveal new gendered power brought about by a digital urban age. As one of the women said when she came to see the exhibition:

> I am really loving this day very much. I feel so happy, I cannot imagine what I have achieved, I never thought I would be able to do this. But I accepted this challenge and completed this. I learnt so much as well. This is only an example, a trailer, the whole film is yet to be done. We have just touched the tip. More power to you sister. [Interview with participant, 2019]

Literally translated from Hindi, *aana jaana* means "coming going". Its meaning and use as a cultural metaphor, however, extends far beyond this literal translation. *Aana* can mean "reverting" or "to occur". *Jaana* can mean "to pass" or "to cease to exist". Aana jaana is a metaphor for journeys, communications, connections, associations, interceptions, social networks, and individual/collective behaviours. We created the hashtag #AnnaJaana to also represent how gendered power is normalized and routinized over women's bodies and spaces. It provokes us to think what *coming going* means in a context where social media provides real time information on the dangers and freedoms located in the metro, bus, auto rickshaw, and walkways, as well as the opportunity to express this in creative and poignant ways.

In this chapter, I examine how gendered life in the margins of social, physical and online spaces is reconfigured in India's "digital turn" (Datta 2018). I ask what happens when women "see" and "speak" through their phone, and how that shapes curation of the city. While so far, much of the practices of "seeing" have been debated through the eyes of the state (Corbridge 2005), and gender empowerment is often seen through the lens of "choice" (Kabeer 1999) or "speech acts", I argue that in the context of limited choice or speech, the possession and use of the mobile phone among poor urban women opens up a new space of participation that cannot be fully captured within these frameworks. Much as Isin and Ruppert (2015) suggest, theirs is an act of saying things by doing things through witnessing, curating, and speaking about violence, and sharing this through technology. I frame this practice as #AanaJaana by suggesting that women living in the urban margins negotiate the freedoms of moving (aana) in online space with the dangers of going out (jaana) into the city, or the restrictions of entering (aana) online space with the freedom of leaving (jaana) home. Through these four moments I draw attention to this metaphor also as a form of networked violence – a set of crosscutting power and control over women's bodies that unfolds with/by the mobile phone and across digital, physical, and social spaces of the home, family, neighbourhood, and city.

#AanaJaana as gendering "networked violence"

#AanaJaana positions itself across three important and distinct debates which so far have remained disconnected. First, it is situated within the prolific debates on violence against women (VAW) and resultant interventions towards safe cities. Planning literature has long charted how urban policies, laws and their implementation have tended to exclude women from both public sphere and public space (Adur and Jha 2018; Viqar 2018; Whitzman, Andrew, and Viswanath 2014). Moreover, beyond planning norms, women's exclusion from the city is reinforced through representation in cultural signifiers such as signage, hoardings and advertising. As Butcher argues in the case of Delhi, these "have created a city of threat and discomfort that problematizes women's access, be it for livelihood or leisure, enclosing women within coordinates not of their making" (2018, 727). For poor women, living in the city is more than mobile place-making (Jirón, Imilan, and Iturra 2016) or quiescence (Bissell 2009); rather their journeys in and out of the city (for work, livelihood or leisure) are undertaken despite the normalization and routinization of violence in all its structural, symbolic and material forms. Although VAW has received sustained attention in Safe Cities policies (UN Women 2017), these are somewhat universalist in approaching "women" as a singular identity whilst dealing mainly with violence in public places. Such intersectional limitations are evident in movements such as "why loiter" (Phadke, Khan, and Ranade 2011) or "meet to sleep" which seeks to reclaim urban public spaces through women's casual presence (akin to a female flâneur). Yet, for poor women in the margins, loitering is not possible given their "time-burdens" (Chant 2013), which restrain their involvement in any activity involving non-productive labour. For women in the margins of class, caste, or other social affiliations, the city is largely experienced through their need to access livelihoods (Tacoli and Satterthwaite 2013).

On the other hand, while violence in online spaces is on the rise, there has been little if any connections made between planning policies, safe cities campaigns, and cyberbullying. Scholars have argued for a long time that the internet is not merely a digital entity that acts as an interface, but as something that has a socio-geographical presence (Rangaswamy and Arora 2015). Rich scholarship on feminist digital geographies (Elwood and Leszczynski 2018) and data feminism (Lupton 2020), now address the multiple ways that gender, sexuality, and race are implicated in the ways that data is collected, stored and shared. Apart from this scholarship, which is situated almost exclusively in the Western context, there is also a dearth of work on how violence circulates in complex ways across physical, social, and online spaces in the context of the global south. I argue here that in a digital age, violence crosses the boundaries of urban public space and the digital public sphere locating poor urban women in the "networked margins" – a "blending together of the edge and the margin" brought about through the proliferation of mobile communication systems amongst the poor while keeping them marginalized through social and structural conditions of oppression (Shah 2015, 9). This paradox is particularly evident in online feminist movements and hashtag campaigns against violence in India and across the world (Subramanian 2015) such as #everydaysexism and more recently #metoo (Bowles Eagle 2015). Women's personal and oral accounts have begun to acquire their own legitimacy through circulation and addition, bringing intersectionality and networks of violence into sharp focus through digital media (Keller, Mendes, and Ringrose 2018). Yet it is safe to say that urban poor women have severely "restricted agency" (Thi Hoan, Chib, and Mahalingam 2016) in adding to these campaigns or even to bear witness to acts of violence as middle-class women. For those living in slums, informal settlements or resettlement colonies, this affirms their double exclusion from both urban and digital life.

The second debate that emerges from the above is situated more specifically on the gendered use of mobile phones, social media and digital technologies amongst the poor and marginalized in the global south. Although scholarship on Information and Communications Technologies for Development (ICT4D) and new media have long critiqued the policy conflation of mobile phones with empowerment (Schech 2002), recent scholarship provides much more nuance and sophistication to this argument stressing that the "digital divide" (Rao 2005) is also a gender divide (Kleine and Poveda 2016). As Gurumurthy, Chami, and Thomas (2016) note, women's access to digital technology is not a sign of empowerment. Rather, we need to consider why despite access to technology, poor women's digital capacity continues to remain at low levels compared to middle class women. Qiu further introduces the idea that the unevenness of distribution of ICT resources is "[serving] as a new seedbed for class formation" (2009, 8). Scholarship on the intersectionality of the digital divide thus argue that mobile phones have not been harnessed sufficiently by historically disadvantaged communities to "empower" themselves, leading to class, caste, and religious stratifications (Kamath 2018; Sarkar 2016). Yet despite these inequalities, mobile phones nonetheless can and do give disadvantaged communities access to more intimate and personal forms of communication that can be transformative (Archambault 2011). The choice of apps to download on mobile phones, their use and personal data shared on these makes them what Poggiali calls a "bridge to people and information worlds away, through free messaging services including WhatsApp and social media sites such as Facebook" (2017, 255).

Willems thus notes that instead of an internet of things, we should consider a "politics of things", examining how "things, objects, infrastructures, and physical space remain crucial to political communication in a digital age as well as to the manner in which bodies, objects, and urban space become politicized and digitally remediated" (2019, 1). Thinking through the politics of mobile phones means addressing its somewhat marginal status in the burgeoning literature on smart cities which has tended to focus on algorithms, platforms, and data production, rather than the embodied nature of the everyday technological objects used by urban poor. Mobile phones produce spaces of "perpetual contact" (Katz and Aakhus 2002), even for those women excluded from urban public spaces. As Archambault notes, mobile phones embody "the potential to shift, albeit in contested and imperfect ways, the interface between daydreams and reality, between mental trips and trips in the material world, by expanding the possible in more palpable ways" (2017, 408). This is relevant in a context of gender-based violence where women face dangers across domestic and public spaces as well as the digital public sphere, yet continue to use mobile phones as "active agents in facilitating their aspiration for individual changes, autonomy, and more powerful decision-making roles in domestic and social domains" (Thi Hoan, Chib, and Mahalingam 2016, 1). For young women whose access to public spaces is controlled by patriarchal norms around gender mobility and employment for example, access to mobile phones provides entry to an online public sphere of social media that recalibrates their sense of belonging from their immediate home and neighbourhood to a geopolitical scale (Carmody 2013). In this context, the widespread use of ICT by young women is an embodiment of a very important socio-cultural transformation taking place in the Indian sub-continent (Punathambekar and Mohan 2019; Rangaswamy 2013; Rangaswamy and Arora 2015).

Finally, the chapter brings the above debates together to contribute to a creative praxis of co-curating the city. While curation emerged in the context of museums, curating the city has been largely confined to the realm of architects, historians or other authorial figures, as a professional act associated with high culture. In a digital age, the proliferation of curatorial platforms such as Pinterest or Instagram means that ordinary citizens can now capture, categorize, narrate, augment and annotate their embodied experiences in the city using just a mobile phone. Communication platforms such as Facebook use algorithms that curate and capture information that can be mined for commercial profit. Nevertheless these platforms also have a sociality – they capture interpersonal relations of friendship, affect, trust, and solidarity – relations that are the basis of traversing across personal, political and public spaces. They enable bottom-up uses of "media technologies as cultural interfaces" (Verhoeff and Wilmott 2016, 119) thus disrupting the distinction between a curatorial object and the curator as *auteur*.

Co-curation in this context involves marginal gendered citizens in "documenting-with" as well as "thinking-with" the mobile phone around the endless curatorial possibilities that it has opened up for them. The mobile phone not only "shapes our visual practices: in the way we act, experience and think with mobility" (Verhoeff and Wilmott 2016, 119), it also shapes the creative practices we use to experience, understand and communicate it to each other. This form of digital curation suggests a shift from a "gatekeeping model to an open model steeped in digital relationships across global networks and the Internet" (Giannini and Bowen 2016). In this practice, the aim is not necessarily to produce works of "Art", but rather to document everyday life as an art form to initiate awareness and advocacy around violence against women. The mobile phone enables its users to become digital curators giving voice and narration to their own experiences within networks of violence across physical, digital and social spaces. Navigating and using curatorial devices such as mobile phones produces a mediated city that brings forth important questions around cultural production, self-representation, consumption, aesthetics, and authorship by marginal social groups. This curation is always incomplete as the vast amounts of data are continually user generated and edited, and WhatsApp messages and Facebook posts are deleted, revised, forwarded, and shared. The data is ephemeral as it stays in cloud servers even if they are deleted from mobile devices, but the curation has endless possibilities that are contingent upon how users want to tell their stories. When this data materializes in a physical exhibition in a public place, it also captures the ethos of the exhibition as #AnnaJaana – of moving between physical and online spaces, not just by women in the margins but also of the digital objects and cultural artefacts that they produce, and therefore represents a materialization of their stories of networked violence across these spaces.

In #AanaJaana, we curated and materialized the ephemerality of networked violence across digital, physical and social spaces that was created and curated by and with women from the margins. The curation presents the "daily condition of women's lives and their lived experience of the technology within a constrained social structure", as a creative praxis that prioritizes a "situated agency" to investigate the relation between mobile phone use and women's voice in this context of violence (Thi Hoan, Chib, and Mahalingam 2016, 2–3). It crafts a whole host of relations with place through the frame of the front-facing camera and built-in microphone, through multiple language keyboards, emojis, as well as a range of tactile positionings that cut across the personal and social, public and private, home and city as well as offline and online spaces. Here the exhibition becomes both a praxis of technology connected to the algorithms and datafication of platforms and apps and simultaneously a craft of graphic design that is made to fit the physical constraints of lightboxes and wall spaces of the concourse in Mandi

House metro station. Its curation is a highly personal and intimate practice that makes the production of the exhibition an empowering experience of self-transformation for those in the margins. The digital object and the exhibition become significant as a connected metaphor, because as Bell notes, it "may also become a form of juris-writing, a writing that concerns and aims at Justice" (2016, 137).

WhatsApp ethnography

The research that informs this chapter began in 2017, when I conducted semi-structured interviews and mental mapping with over thirty women in a slum resettlement colony in South Delhi. Through subsequent research funding we recruited twelve women from amongst the earlier group to participate in an in-depth and extended research into gendering the top-down vision of a smart city. As we began this phase of work, we immediately came across a set of challenges around the distances of geographical time and space between researchers in the West and participants in India. Because we wanted to examine their digital lives, it was also not possible to use conventional methods of ethnography and interviews. We agreed to set up a closed WhatsApp group that would instead enable real-time free flowing conversations that could somewhat bypass the challenges of physical distance. We therefore requested WhatsApp diary entries from participants to document their daily experience of the city. While this addressed the challenges of arranging regular face-to-face interviews or mobile observations of their journeys in/out of the city, we also simultaneously scheduled face to face workshops between participants and the research team during key stages of the project. This closed WhatsApp group was the start of what developed into a "WhatsApp ethnography" over six months.

India has the highest number of WhatsApp users in the world today, closely followed by Brazil. The popularity of the app in the country crossed 15 million active users in 2014, a year after the company dropped its subscription model (charging users a dollar per year after first year of use). Its easy usability, inbuilt characteristics to make private groups and ad-free nature makes it a highly popular medium across all ages. This is specifically the case amongst the young in India, who have propelled the information revolution that is taking place, especially since the availability of cheap smart phones and cheaper data packages. WhatsApp has continued to be the most popular, widely used social messaging app in

India over the last five years. For us to be able to set up a closed WhatsApp group for research purposes was speaking to these transformations in digital communication, especially in the lives of the young millennials in a country like India.

Their WhatsApp entries constituted an "everyday dwelling" which were "constitutive of a felt-life of being together with those close by" (O'Hara et al. 2014, 1). Curation with WhatsApp occupies a praxis and politics of mobile phones, both in their embodiment *of* experience as well as in their tools of digital documentation *as* experience. Initially the WhatsApp diaries consisted of loose and irregular entries – a shortened text message, captioned photo, or an audio file, and sometimes even a video. These would illustrate the positive or negative aspects of their journeys to the city and back home. The absence of locative media within WhatsApp also meant that visual and aural diary entries had to be enriched by textualization. Since we had already conducted in-depth interviews with these women, we could contextualize most of these entries without extensive conversations about them. Eventually, the entries began to show more structure and pattern – there was a flurry of entries during festivals or during particularly difficult moments in their daily commute such as monsoon rains that led to traffic congestion and delayed their journeys. There were also several entries highlighting the prevalence of violence – whether gender-based or otherwise.

Slowly a dynamic relationship began to emerge within the group with some of the participants becoming more regular contributors with others remaining quieter and needing more encouragement to send in their entries. These power relationships diversified over time, when a few participants began to post fake news or forwarded messages while others called them out. In that sense, our small, closed WhatsApp group became a reflection of wider power dynamics in society. WhatsApp could not totally replace the corporeality of face-to-face contact, which was still maintained from time to time, but the WhatsApp entries made moments of physical contact more meaningful and exciting as we recounted and reflected upon them. The WhatsApp group itself became the site of knowledge exchange and co-production, with participants sometimes putting up pleas for jobs opportunities, or circulating information on government schemes, or organizing Women's Day marches, or watching out for each other and sometimes even falling out through disagreements.

It was during one of our workshop sessions that our research assistant said that she had noticed that these interactions were all about the participants' aana jaana. This observation was reinforced when after one of the WhatsApp photo entries we asked, "What does this show?" The participant promptly replied, “It shows the aana jaana of people in the city.” The notion of aana jaana came up repeatedly in the diary entries and led us on an intellectual and curatorial journey into a colloquial cultural metaphor which stood for the visual and digital lives of women in Delhi’s urban peripheries. In this way, curation of the city emerged organically in our group – first in the discussions on why participants curated particular experiences as digitally significant enough to upload on the group as images, audio or text, and second when they participated in identifying the theme of the exhibition as #AanaJaana and provided some feedback on the kinds of entries that could be included in the physical space. The final physical format of the exhibition however was undertaken directly by a graphic artist who created this in collaboration with the researchers and project partners.

As researchers, we had to continually reflect upon the ethics of using WhatsApp with often vulnerable young women. Globally, there is an emerging research interest with or on WhatsApp (Ahad and Lim 2014; Dixon 2018; O’Hara et al. 2014). Misinformation shared through WhatsApp has led to large-scale violence and elicited many campaigns. The fact that this was a “new” kind of online ethnographic space was both exciting and came with an added sense of responsibility, especially when conversations on ethics on digital methodologies are still ongoing (Heeks 2008). We followed Winter and Lavis, who note an online ethnography of “listening as a mode of participating in, as well as observing, online spaces” (2019, 55). This method enabled diverse interactions, spread across time and space, between their temporal spaces in Khadar, as well as between themselves and the authors across geographies and continents. We were wary of the possibility of excess circulation or leakages, when digital content, specifically images, get forwarded into online networks where they take a life of its own. This consciousness continuously determined our interactions. However, this concern often collapsed when awareness was shown by the participants themselves with an alert disposition to these aspects. While we had more access to different mediums and information from the web which could be used in the conversations, WhatsApp also somewhat suppressed the physical embodiment of class or caste backgrounds in the conversations as we all became digital texts, using the same languages and idioms (emojis or shorthand) that was enabled by the platform. This absence of a face allowed for some seamless communication – the medium and its construction as a social media app with easily accessible limited features afforded everyone a sense of equalizing space.

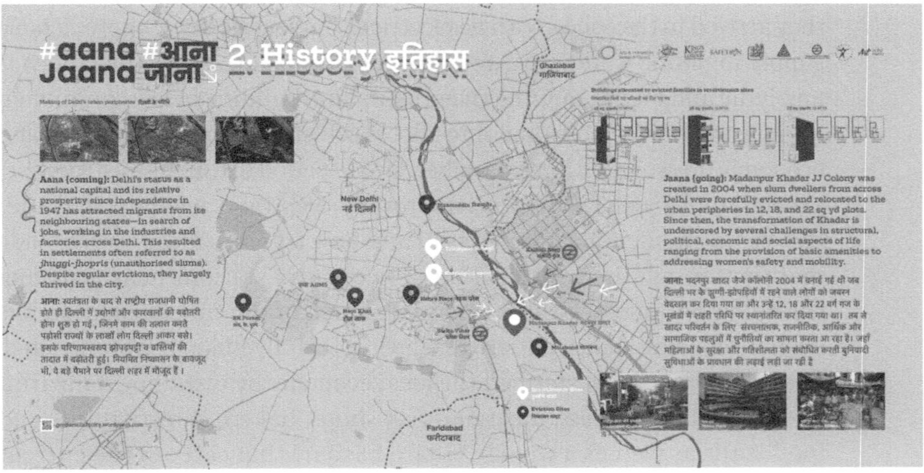

Figure 4.2. The history of slum evictions in Delhi with location of Madanpur Khadar in the south-eastern periphery. Courtesy of Kruttika Susarla.

Madanpur Khadar JJ colony: making of the urban periphery

All the young women participants lived in one of Delhi's slum resettlement colonies on its periphery. This physical location was part of the migrant itineraries of participants' families – i.e., they evoked aana jaana – a coming and going that is both historical and geographical. Rural migrants who came to ride the city's prosperity through the 1960s, 1970s and 1980s found affordable housing in its many slum and informal settlements, but since the late 1970s have been forcefully evicted in waves of city beautification schemes (Tarlo 2003). Madanpur Khadar JJ colony was created after a judicial ruling in the early 2000s radically transformed the identities of Delhi's slum dwellers into "pickpockets of urban land", as illegal citizens (Datta 2012) that then led to a spate of slum demolitions across the city. This literally "swept off the map" (Bhan and Menon-Sen 2008) all material evidence of slums from Delhi's cartographic representation to then relocate them at the outskirts of the city.

Madanpur Khadar JJ Colony is a space of double exceptionality – first by being created without history and context close to the villages and border towns that were antagonistic about the introduction of "outsiders" within local social and kinship networks; second by being "urbanized" through their inclusion into Delhi's formal planning structure, within typologies of incremental building and formal access to urban basic services, that are still denied to the villagers. Residents however faced many challenges when they arrived in the 2000s, with no electricity, no sewage lines, unpredictable supply in public water taps, and no public transport. Since the land was low-lying and on the Yamuna river floodplain, building bylaws prohibited the construction of more than three floors, which often put severe space and economic pressures on larger families. Over a period of time, infrastructural, economic and social conditions improved – electricity was provided after five years, public toilets were built, and new transport links were made mainly through the construction of two metro stations nearby during the 2010 Commonwealth Games. Still, much of the transformation has been through residents' initiatives who lost their livelihoods when evicted; nonetheless they transformed the local economy by installing daily fruit and vegetable markets, meat and fish shops, bakeries, restaurants and food stalls as well as a weekly "Shani bazaar" that cater to the current resident population of 25,000. It is also through residents' initiatives that the municipality has been forced to recognize their status and provide better sanitation, drainage facilities and water supply. Recently the youth have also begun to make vehement claims for Delhi Transport Corporation buses to provide direct connections from Khadar to the nearby metro stations, a journey which is otherwise fraught with the uncertainties and dangers of privately run e-rickshaws. Physical connectivity is of crucial importance to residents in Khadar since most residents are in low-wage jobs as peons, drivers, domestic workers, security guards, rag pickers, vendors, and construction, industrial and commercial workers across the city. Young women on the other hand are engaged mostly in service sector work across the city as factory workers, low end Business Process Outsourcing, taxi drivers, and NGO workers.

#AanaJaana: curating the city with mobile phones

What does it mean to curate #AanaJaana in a context where there are no longer clearly defined boundaries between the physical and the digital? The women participants in our study were digital natives embodying the paradoxical nature of the networked margins. Physical infrastructure in the peripheries might be absent, broken, or reflect the slow time of service (water, sewage, sanitation, electricity, and broadband fibre-optic cables), but connectivity via mobile telephones apparently offered them access to information and knowledge without material improvements in urban basic services. At the same time, the promise of connectivity and communication embodied in the phone also remained unfulfilled with the lack of

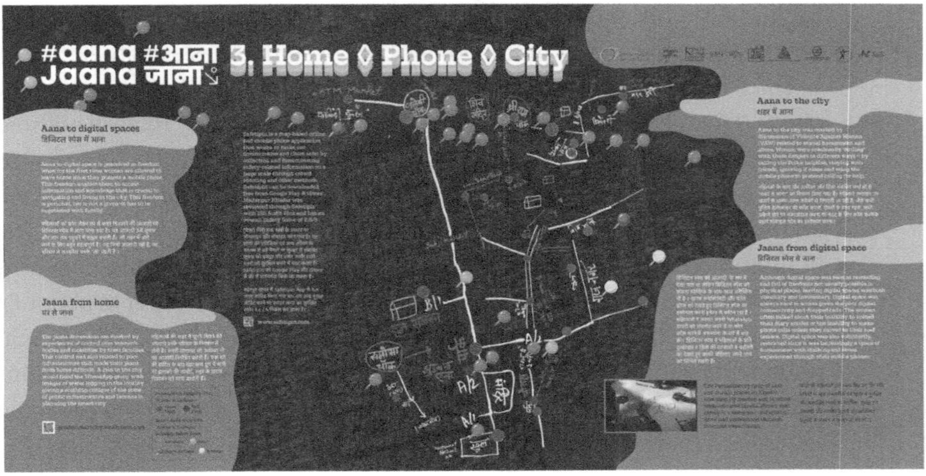

Figure 4.3. Four propositions of #AanaJaana. Exhibition in Mandi House Metro station, Delhi. Courtesy of Kruttika Susarla.

network coverage in the peripheries, dropped calls, older models of mobile phones which crashed apps, slow network speeds, and so on. Thus, while the women were relatively better educated, more digitally connected and more employable when compared to earlier generations, they were less able to exercise these freedoms because of constraints on social, digital and physical infrastructures.

In our exhibition, we drew upon this notion of #AanaJaana to present four propositions – aana to digital space, jaana from home, aana to the city and jaana from digital space. These curate four moments of the paradox of inclusion and exclusion, belonging and alienation that highlight the networked violence of gendered digital lives in Delhi.

Aana to digital space: confined freedoms

Coming into digital space was often fraught with negotiations of material conditions for young women. In the absence of laptops or desktops, participants accessed and imagined digital space primarily through the mobile phone. The phone was seen as an aspirational material possession, as well as a tool of empowerment which would connect them to people and networks that they had so far been left out of. Participants claimed they had to negotiate hard with their families to own a mobile phone, particularly if they were married and were living with in-laws. They argued that they needed the phone to stay safe and keep in touch while they went out for "legitimate" work in the city. Often, they would be given the phone owned by a family member – their mother or brother, or sometimes even their father.

Once they started earning, their first salary was usually spent on a new phone. They chose a cheaper model of Samsung, Xiaomi, or Huawei android phone, costing between Rs 5500-7500 (£55-£75) which had the capacity to provide them access to different apps – YouTube, Facebook, WhatsApp, and so on. For some, this was more than their monthly earnings, but the proliferation of mobile phone shops offering phones by instalment made it possible to get one of their preferred brands. Data was cheaper at about £5 a month for 2GB, and connectivity was almost instantaneous if they could provide ID verification. Through these social and financial negotiations and compromises, the phone became the conduit to digital and personal space, and therefore to freedom and safety:

> [I]f a girl is alone, if she has any weapon – she has a phone. Even at the last moment, she can call someone, with 100 or some urgent call. You can call home and inform them and ask someone to come, urgent, please come pick me. [Interview with participant, 2018]

Yet, as I suggest later, this notion of freedom and safety was often more imagined than real. In everyday commutes to the city and back as well as within their neighbourhood, participants constantly faced cat calls, molestation and other forms of misogyny. Despite their capacity to connect to wider digital worlds, this violence also extended to the closed spaces of chatrooms such as WhatsApp and social media such as Facebook. This led to the women learning how to initiate stronger privacy controls after breaches where they were harassed or subject to online abuse.

However, aana to digital space was a freedom that is nonetheless confined to closed spaces of WhatsApp groups or Facebook. It was not common for these women to use internet browsers such as Google Chrome or Internet Explorer. Instead, much of their information was mined from unsolicited messages from third parties and WhatsApp forwards. This mean that often they would fall for propaganda material and circulate this to other networks, or fall prey to fake news around terrorism, nationalism or religion. This closed freedom was what informed and constantly determined their access to current affairs and the opinions they developed on contemporary politics. This to us implied that in order to really deepen the notion of ICT based empowerment, it is important to look beyond just access and engage with ways to break the closed social enclaves that permeate into their digital experience as well. The paradox lay in the fact that the digital field opened to them an access to developments around the #metoo movement, as well as be recipients of numerous amounts of hate-speech and misinformation campaigns spearheaded through WhatsApp.

Jaana from home: infrastructure curfews

One of the common features of their aanajaana was the curfew imposed by their families. The jaana discussions are marked by experiences of control over women's bodies and mobilities by their families.

> If you take an average girl who is going at 9 in the morning and returning at 6 [...] if she is delayed and it is 6:30 instead of 6, many times they are beaten up by their parents. 'Where were they for half an hour! Who were you roaming around with!?' And would give really degrading verbal abuses. [Interview with participant, 2018]

The women were clear that the curfew as a way of coping with imminent danger revealed the fault lines of intergenerational relationships – its asynchronicity between them as "digital natives" and the older generation who have grown up without mobile technologies and cannot understand that the curfew does not necessarily keep them safe from online abuse and sexual harassment. They were conflicted between the need for more security and surveillance, for example increased street lighting, police presence, and CCTV cameras, and the realisation that these did not keep men or women completely safe:

> Earlier in our neighbourhood, there were no cameras. In the Baraat Ghar [Community Hall], many incidents used to take place, one or two girls have been raped. Who raped whom, we were never sure. But now, there are cameras installed there, so everything is watched, whoever is coming and going, and the inside everyone is watched, what are the boys up to? [Interview with participant, 2018]

Yet while they acknowledged that fixing CCTV cameras might result in identifying criminals after the crime, they were also aware that no one is safe in Khadar. This emerged in the conversations in the aftermath of a murder that took place on the main street, which is a very crowded busy artery leading into the colony. This is a daily market, typical of these colonies that bustle with shops selling snacks, recharge coupons and groceries. And

Figure 4.4. WhatsApp diary entries on safety. Courtesy of Kruttika Susarla.

here the murder of a young man occurred under the watchful eyes of the CCTV camera. There was fear, which was palpable, and which was shared constantly on our group. These conversations even led to community-based organizations following up on the case, conducting a safety walk in the community, auditing and marking out the darker spots within the resettlement colony. It further initiated conversations on increasing violence and the need to address it collectively, as a community of inhabitants of the colony. This generated several images of the man's body and newspaper cuttings reporting the stabbing, as well as WhatsApp entries about safety in their neighbourhood in general:

> We should do something in Khadar. It's becoming scary to stay here, and these days it's scary to say anything to anyone! [WhatsApp diary entry, 2018]

Jaana from home was also related to poor physical infrastructures of drainage and transport that made their daily commutes to work or college difficult. Between June and August, our WhatsApp group was flooded with images of waterlogging in the neighbourhood accompanied by scathing critiques of the state of public infrastructure and lacunas in urban planning:

> Today while returning from office in a bus, I got stuck in a traffic jam, so much that the route which takes 10 minutes to cover, it took me 1.5 hours to get back. I got on the bus at 4.15 and got out at 5.30. The jam was because of the rainwater flooding the street; there was so much water that only the handles of the bikes were visible, and one bike in front of me fell into the water. The bus was moving so slow, I felt impatient and felt like getting out, but because of water, I didn't. [WhatsApp diary entry, 2018]

These blockades in their commutes that were both daily (traffic) and seasonal (monsoons) contributed to a wider feeling of being under infrastructural curfew. These restrictions were both social and physical and reinforced the overall sense of "immobility" felt by the women.

Aana to the city: selfies as geotagging

Like the transformative potential of mobile phones which scholars have examined in African countries (Archambault 2011; Carmody 2013), one of the key "freedoms" of owning a mobile phone while "coming" to the city

Figure 4.5. WhatsApp diary entries on infrastructure. Courtesy of Kruttika Susarla.

is the ability to express oneself in ways not possible before. While taking photographs was not common amongst their families (unless for passports or weddings), the mobile phone enabled the rise of much more candid urban photography than existed before.

Our participants became avid selfie makers when they went into the city. These selfies were not only for the "wild and everyday", as Rangaswamy and Arora (2015) describe digital leisure in Indian slums; they were also an analogue format of "geotagging" their location in the city. Selfies were rarely taken inside the home; they were primarily taken outside, when they got dressed up, and were with their friends. These selfies represented their identities as young urban women travelling on the metro or bus, going out with friends, eating out or going to women's day marches. Right after our workshops, our phones were inundated with innumerable selfies with us and the places where these workshops were held. Clearly the workshops were a culmination of their solidarity and friendship with us as researchers. They curated the city through selfies – any place they visited would be immediately documented with them in the foreground and the city in the background. These selfies represented a framing of the gendered self as well groomed, smart and well located. And they tagged the city literally at arm's length, staging it as a backdrop in the front facing phone cameras. The selfies represented their aana – their coming of age in the city, framing their friendships, networks, solidarities and expressions of affection for each other.

Selfies were created in a context where networked violence was the pervasive experience of coming to the city. They marked their "freedoms" as small moments during the continuous anticipation of violence during their travelling into and moving through the city. In this context, safety was not seen through the geolocational capacities of Google or other safety apps but rather in using the mobile phone for instantaneous communication with their family and friends. The mobile phone thus became the curator of violence experienced and often visualized through selfies in the city:

> I had to go from Jasola School to Badarpur village to my office just now. I wasn't able to get a bus, and then suddenly a private bus came which was headed for the border. The moment I got inside I saw there were only men inside. I got scared because everyone was staring at me. [WhatsApp diary entry accompanied by selfie, 2018]

Figure 4.6. WhatsApp diary entries of selfies. Courtesy of Kruttika Susarla.

Jaana from digital space: communication blackout

Jaana from digital spaces was both voluntary and involuntary. Digital space was sometimes voluntarily restricted since it was increasingly a space of harassment, with stalking and abuse experienced through their mobile phones:

> So, I was in the minibus once. Somebody was messaging this girl, saying very random things, I shouldn't be seeing it, but my eyes fell on it, so they were exchanging – who are you etc., and it reached a level where they were abusing each other on it. So, I said, sorry, I saw your message, but haven't you put privacy? She asked, how to? So, I explained to her that there are privacy settings in WhatsApp. [interview with participant, 2018]

Digital space was always hard to access given poor digital connectivity and dropped calls. Women often talked about their inability to upload their diary entries or the inability to make phone calls unless they moved to their roof terrace. A picture of a subway without any light in the middle of the day elicited much concern and conversation on our group. One of the participants also shared that the phone network is often interrupted while inside the subways. This connected different networks of communication to each other even as women struggled to make time to climb flights of stairs up to the rooftop of their houses:

> I am not able to send any audio because my phone has hung and it needs fixing, I'm sending from friend's phone. [WhatsApp diary entry, 2018]

Speaking back with "things", but will the city listen?

This chapter has presented the contradictions of speaking with "things" when structural and social conditions of exclusion are historic and persistent across public and private spaces. By using a medium such as WhatsApp diaries we were able to provide a "safe space" to have open and supportive conversations that were in themselves transformative and creative. This produced a new kind of politics around the use of technology. #AanaJaana highlighted the contextual use of technology in self-expression and thinking while doing, that went beyond just "empowerment". Although the mobile phone has become a key theme in research on young people, #AanaJaana highlights that the mobile phone is also instrumental to how young women

Figure 4.7. WhatsApp diary entries on challenges with digital connectivity in exhibition panel. Courtesy of Kruttika Susarla.

experienced exclusionary politics across different spaces of home, city and digital public sphere. It produced new understandings of safety, subjectivity and identity and it showed how the phone played an important role in cultivating an intimate sense of power/lessness and ways of creatively engaging with these through an exhibition.

As one of the participants said at the start of this chapter, this exhibition was only a "trailer". Its afterlife emerged in the multiple ways that issues and themes raised in the exhibition began to contour public campaigns in the city. Beyond its very public coverage by national media channels which gave voice and space for these women to speak, it provoked a public discourse in the city through calls to include violence against women as an agenda in upcoming local elections. Further, these women were invited to join ongoing consultations organized by relevant NGOs towards a new gender-sensitive masterplan for Delhi. Co-curation provided a space to change the terms and conditions of marginal digital lives and their participation in the public sphere if only by drawing attention to the nodes of networked violence kept invisible and normalized in the peripheries. It brought safety centre stage, into public visibility in a metro station which sees an average daily footfall of 11350 commuters. The exhibition and its afterlife suggest that we need to engage much more in critically thinking of the role of technological "things" such as mobile phones and popular platforms such as WhatsApp that are used by those in the margins and find ways to use these as tools for transformative change. This cannot be arranged easily and indeed our work was limited to a small group of people, but their motivation in taking part in this "trailer" was driven by the need to tell their story to others like them. This motivation has the potential to lay the ground for a new kind of creative and critical movement against networked violence led by those still on the margins of digital and urban life.

References

Adur, Shweta M., and Shreyasi Jha. 2018. "(Re)Centering Street Harassment – an Appraisal of Safe Cities Global Initiative in Delhi, India." *Journal of Gender Studies* 27 (1): 114–24.

Ahad, Annie Dayani, and Syamimi Md Ariff Lim. 2014. "Convenience or Nuisance?: The 'WhatsApp' Dilemma." *Procedia — Social and Behavioral Sciences* 155: 189–96.

Archambault, Julie Soleil. 2011. "Breaking up 'Because of the Phone' and the Transformative Potential of Information in Southern Mozambique." *New Media & Society* 13 (3): 444–56.

———. 2017. *Mobile Secrets: Youth, Intimacy, and the Politics of Pretence in Mozambique*. Chicago: University of Chicago Press.

Bell, Vikki. 2016. "Between Documentality and Imagination: Five Theses on Curating the Violent Past." *Memory Studies* 11 (2): 137–55.

Bhan, Gautam, and Kalyani Menon-Sen. 2008. *Swept off the Map: Surviving Eviction and Resettlement in Delhi*. New Delhi: Yoda Press.

Bissell, David. 2009. "Travelling Vulnerabilities: Mobile Timespaces of Quiescence." *Cultural Geographies*, October.

Bowles Eagle, Ryan. 2015. "Loitering, Lingering, Hashtagging: Women Reclaiming Public Space Via #BoardtheBus, #StopStreetHarassment, and the #EverydaySexism Project." *Feminist Media Studies* 15 (2): 350–53. https://doi.org/10.1080/14680777.2015.1008748/.

Butcher, Melissa. 2018. "Defying Delhi's Enclosures: Strategies for Managing a Difficult City." *Gender, Place & Culture* 25 (5): 727–42.

Carmody, Pádraig. 2013. "A Knowledge Economy or an Information Society in Africa? Thintegration and the Mobile Phone Revolution." *Information Technology for Development* 19 (1): 24–39.

Chant, Sylvia. 2013. "Cities through a 'Gender Lens': A Golden 'Urban Age' for Women in the Global South?" *Environment and Urbanization* 25 (1): 9–29. https://doi.org/10.1177/0956247813477809.

Corbridge, Stuart. 2005. "Governance." In *Seeing the State: Governance and Governmentality in India.*, edited by Stuart Corbridge, Glyn Williams, M. K. Srivastava, and Rene Veron, 151–87. Cambridge: Cambridge University Press.

Datta, Ayona. 2012. *The Illegal City: Space, Law and Gender in a Delhi Squatter Settlement*. Farnham: Ashgate.

———. 2018. "The 'Digital Turn' in Postcolonial Urbanism: Smart Citizenship in the Making of India's 100 Smart Cities." *Transactions of the Institute of British Geographers* 43 (3): 405–19.

Datta, Ayona, and Arya Thomas. 2021. "Curating #AanaJaana [#ComingGoing]: Gendered Authorship in the 'Contact Zone' of Delhi's Digital and Urban Margins." *Cultural Geographies*, https://doi.org/10.1177/1474474021993415/.

Dixon, Natalie. 2018. "Stranger-Ness and Belonging in a Neighbourhood WhatsApp Group." *Open Cultural Studies* 1 (1): 493–503.

Elwood, Sarah, and Agnieszka Leszczynski. 2018. "Feminist Digital Geographies." *Gender, Place & Culture* 25 (5): 629–44.

Giannini, Tula, and Jonathan P Bowen. 2016. "Curating Digital Life and Culture: Art and Information." In *Electronic Visualisation and the Arts (EVA)*. http://dx.doi.org/10.14236/ewic/EVA2016.46/.

Gurumurthy, Anita, Nandini Chami, and Sanjana Thomas. 2016. "Unpacking Digital India: A Feminist Commentary on Policy Agendas in the Digital Moment." *Journal of Information Policy* 6: 371–402.

Heeks, Richard. 2008. "ICT4D 2.0: The Next Phase of Applying ICT for International Development." *Computer* 41 (6): 26–33.

Isin, Engin, and Evelyn Ruppert. 2015. *Being Digital Citizens*. London: Rowman & Littlefield.

Jirón, Paola A, Walter Alejandro Imilan, and Luis Iturra. 2016. "Relearning to Travel in Santiago: The Importance of Mobile Place-Making and Travelling Know-How." *Cultural Geographies* 23 (4): 599–614.

Kabeer, Naila. 1999. "Resources, Agency, Achievements: Reflections on the Measurement of Women's Empowerment." *Development and Change* 30 (3): 435–64.

Kamath, Anant. 2018. "'Untouchable' Cellphones? Old Caste Exclusions and New Digital Divides in Peri-Urban Bangalore." *Critical Asian Studies* 50 (3): 375–94.

Katz, J.E., and M. Aakhus. 2002. *Perpetual Contact: Mobile Communication, Private Talk, Public Performance*. Cambridge: Cambridge University Press.

Keller, Jessalynn, Kaitlynn Mendes, and Jessica Ringrose. 2018. "Speaking 'Unspeakable Things': Documenting Digital Feminist Responses to Rape Culture." *Journal of Gender Studies* 27 (1): 22–36.

Kleine, Dorothea, and Sammia Poveda. 2016. "Digital Divide." In *International Encyclopedia of Geography: People, the Earth, Environment and Technology*, edited by Douglas Richardson, Noel Castree, Michael F. Goodchild, Audrey Kobayashi, Weidong Liu, and Richard A. Marston, 1–5, London: John Wiley & Sons.

Lupton, Deborah. 2020. *Data Selves*. Cambridge: Polity Press.

O'Hara, Kenton P., Michael Massimi, Richard Harper, Simon Rubens, and Jessica Morris. 2014. "Everyday Dwelling with WhatsApp." In *Proceedings of the 17th ACM Conference on Computer Supported Cooperative Work & Social Computing*, 1131–43. CSCW '14. New York: ACM. https://doi.org/10.1145/2531602.2531679/.

Phadke, Shilpa, Sameera Khan, and Shilpa Ranade. 2011. *Why Loiter?: Women and Risk on Mumbai Streets*. New Delhi: Penguin Books.

Poggiali, Lisa. 2017. 'Digital Futures and Analogue Pasts? Citizenship and Ethnicity in Techno-Utopian Kenya'. *Africa* 87 (2): 253–77.

Punathambekar, Aswin, and Sriram Mohan. 2019. *Global Digital Cultures: Perspectives from South Asia*. Ann Arbor: University of Michigan Press.

Qiu, J. 2009. *Working-Class Network Society: Communication Technology and the Information Have-Less in Urban China*. Cambridge, MA: MIT Press.

Rangaswamy, Nimmi. 2013. "Digital Life of the Youth in India — What Is the Current Scenario and Where Are We Headed?" *Digital Equality* (blog). 6 December 2013. https://digitalequality.in/internet-magic-in-indian-slums/.

Rangaswamy, Nimmi, and Payal Arora. 2015. "The Mobile Internet in the Wild and Every Day: Digital Leisure in the Slums of Urban India." *International Journal of Cultural Studies* 19 (6): 611–26.

Rao, Siriginidi Subba. 2005. "Bridging Digital Divide: Efforts in India." *Telematics and Informatics*, 22 (4): 361–75.

Sarkar, Sreela. 2016. "Beyond the 'Digital Divide': The 'Computer Girls' of Seelampur." *Feminist Media Studies* 16 (6): 968–83.

Schech, S. 2002. "Wired for Change: The Links between ICTs and Development Discourses." *Journal of International Development* 14 (1): 13–23.

Shah, Nishant. 2015. "Networked Margins: Revisiting Inequality and Intersection." In *Digitally Connected: Global Perspectives on Youth and Digital Media*, edited by Sandra Cortesi and Urs Gasser, 9–11, Berkman Center Research Publication No. 2015-6, https://papers.ssrn.com/sol3/papers.cfm?abstract_id=2585686/.

Subramanian, Sujatha. 2015. "From the Streets to the Web: Looking at Feminist Activism on Social Media." *Economic and Political Weekly* L: 71–78.

Tacoli, Cecilia, and David Satterthwaite. 2013. "Gender and Urban Change." *Environment and Urbanization* 25 (1): 3–8.

Tarlo, Emma (2003) *Unsettling Memories: Narratives of the Emergency in Delhi*. Oakland, CA: University of California Press.

Thi Hoan, Nguyen, Arul Chib, and Ram Mahalingam. 2016. "Mobile Phones and Gender Empowerment: Enactment of "Restricted Agency"." In *ICTD '16: Proceedings of the 8th International Conference on Information and Communication Technologies and Development*, 1–10, Ann Arbor, MI.

UN Women. 2017. "Safe Cities and Safe Public Spaces: Global Results Report." New York: United Nations.

Verhoeff, Nanna, and Clancy Wilmott. 2016. "Curating the City: Urban Interfaces and Locative Media as Experimental Platforms for Cultural Data." In *Code and the City*, edited by Rob Kitchin and Sung-Yueh Perng, 116–129, London: Routledge.

Viqar, Sarwat. 2018. "Women's Public Lives and Sovereign Arrangements in Karachi's Inner City." *Gender, Place & Culture* 25 (3): 416–33.

Whitzman, Carolyn, Caroline Andrew, and Kalpana Viswanath. 2014. "Partnerships for Women's Safety in the City: 'Four Legs for a Good Table'." *Environment and Urbanization* 26 (2): 443–56.

Willems, Wendy. 2019. "'The Politics of Things': Digital Media, Urban Space, and the Materiality of Publics." *Media, Culture and Society* 41 (8): 1–18.

Winter, Rachel, and Anna Lavis. 2019. "Looking, But Not Listening? Theorizing the Practice and Ethics of Online Ethnography." *Journal of Empirical Research on Human Research Ethics* 15 (1–2): 55–62.

About the author

Ayona Datta is Professor in Urban Geography in University College London, UK. Her research interests are in gender citizenship, urban futures and smart cities in the global South. In particular, she is interested in how cities seek to transform themselves through utopian urban visions of the future and their impacts on everyday social, material and gendered geographies. She is author of *The Illegal City: Space, law and gender in a Delhi squatter settlement* (2012), co-editor of *Translocal Geographies: Spaces, Places, Connections* (2011) and *Mega-Urbanization in the Global South: Fast Cities and New Urban Utopias* (2016). Ayona is co-editor of *Urban Geography* journal, on the Trustees' Board of the IJURR Foundation and on the editorial boards of the journals *Antipode, Digital Geography* and *Society and EPD: Society and Space*. For her contributions to understanding of smart cities through fieldwork she received the Busk Medal from Royal Geographical Society (with IBG) in 2019.

5. Future Urban Imaginaries: Placemaking and Digital Visualizations

Monica Degen and Isobel Ward

Abstract

This chapter examines the temporal politics of urban redevelopment by analysing how multiple future urban imaginaries are mediated through a range of digital visualizations and across urban spaces. Focusing on the planned move of the Museum of London into West Smithfield Market in 2024 as part of the Culture Mile redevelopment project, we analyse how a variety of stakeholders mobilize different images across three kinds of spaces – urban space, strategic planning space, and social media space – and how each is underpinned by various and multiple temporalities. We show how digital visual technologies have become central to preparing the ground for urban redevelopment schemes and placemaking strategies, by appealing to our sensory and experiential sensibilities.

Keywords: urban imaginaries, placemaking, futures, Culture Mile

Introduction

Since the 1990s, culture-led regeneration has been at the forefront of many cities' post-industrial revitalization strategies (Cronin and Hetherington 2008; Della Lucia and Trunfio 2018; Miles and Miles 2004; Rius Ulldemolins 2014). Central to this process has been the physical redesign of neighbourhoods linked to a conscious reimagining of the future "look and feel of place" promoted visually through various traditional media outlets such as billboards, magazines and newspapers. Due to changes in media use and advances in technologies during the last two decades, the use of digital visualizations such as computer-generated images (CGI) or sharing digital images on social media has become more prevalent in the promotion of

Rose, G. (ed.), *Seeing the City Digitally. Processing Urban Space and Time.*
Taylor & Francis Group, 2022
DOI 10.5117/9789463727037_CH05

such urban developments. CGIs are used by architects both to design new areas, sell new developments to prospective clients and used on hoardings to promote them (Degen et al. 2017; Rose et al. 2016). Images and films on social media platforms such as Instagram further brand, frame and disseminate the perception and use of these redesigned places (Braun Erik et al. 2013). Greenberg refers to such processes as the production of an "urban imaginary", in other words, the ways in which "the space of the city is produced not only materially and geographically but also in the social imagination and through changing modes of representation" (2000: 228).

This chapter will analyse how shaping and manufacturing this urban imaginary, increasingly based on digital technologies, has become a crucial feature of urban redevelopment schemes and placemaking strategies. In particular, we analyse the ways in which digital visual technologies frame the future senses of place. We show how placemaking is a complex temporal achievement organized and materialized by the interactions between multiple and diverse temporalities framed by urban decision makers, the built environment and its users. We do so by focusing on an area currently undergoing redevelopment: the Smithfield Market area in London. In 2017, the Smithfield Market area was designated part of the Culture Mile[1], London's largest cultural regeneration project for the next decade comprizing fifteen per cent of the total area of the Square Mile, "the citadel of money making" where "creativity is fast becoming the most valuable currency" (Pickford 2017). As part of this project the Museum of London (MoL) is planning to move in 2024 to West Smithfield Market which currently contains derelict Victorian buildings and adjoins the oldest operational wholesale meat market in the UK. Included in the wider redevelopment, is the opening of Farringdon East Crossrail Station in 2021 to improve access to what has been described as London's new cultural destination.

This chapter analyses the relationship between the relocation of the MoL to Smithfield Market and the planning and imagining of the area's future as part of the Culture Mile. The chapter will start by discussing the relationship between placemaking, urban imaginaries and time. We then provide an overview of the Culture Mile and its main aims and objectives and examine how the urban regeneration of the Smithfield Market area and the redesign of the MoL draw on a range of distinct, yet interlocked temporalities. By doing this we want to think through the ways in which the future relocation of the MoL is perceived, imagined and constructed across

1 A partnership endorsed by the Corporation of London including the Barbican, the London Symphony Orchestra, The Guildhall School of Music and Drama and the Museum of London

three distinct spatialities. The spatialities are: a) urban space, b) strategic planning space and c) social media space.

Placemaking, future urban imaginaries and the digital

Since the 1980s, changes in global political and economic processes have promoted a move in urban policy from concerns for welfare issues and social politics to more entrepreneurial strategies where cities are eager to compete for investment and visitors on a global scale. The outcome has been the emergence of a new spatial logic reflected in major urban restructuring, as modernist industrialism has been replaced by post-industrial flexible accumulation (Harvey 1989; Hubbard and Hall 1998; Zukin 1991). These changes are mirrored in the radical redesign of urban landscapes across the globe from Sao Paolo to Manchester which has led "cities across the world to take on a new character and a new dynamic that has forced issues of culture and consumption more predominantly to the fore" (Cronin and Hetherington 2008, 1).

We can understand the increased pressure on cities to brand themselves and promote their unique place-differentiating qualities as part and parcel of this global entrepreneurial arena of urban competition which has encouraged a conscious construction of coherent place identities that will appeal to certain social groups, similar to a commercial product, so called "brandscapes" (Klingmann 2007). Much of this urban branding is now undertaken by formalized public-private coalitions which bring together the competing interests from local government, businesses, real estate and local not-for-profit organizations. Indeed, there has been a clear co-evolution of urban redevelopment and branding, where increasingly the two processes work hand in hand and are part of urban policy "emerging as a hybrid materialization representing the process of creating new spatial settings" (Lucarelli 2018, 12).

This amalgamation of spatial restructuring linked to branding processes is particularly typical for the restructuring of neighbourhoods, rather than a whole city where a symbolic layer is often just added to an existing place, illustrated by slogans such as "Barcelona more than ever" or "I Amsterdam". Instead, neighbourhood regeneration tends to encompass the combination of a redesign of the physical landscape linked to a broader array of urban policies (e.g. event management, retail and leisure infrastructure, zoning policies, and so on) and branding processes which re-signify and aim to choreograph the sensory-emotional experiences or, to put it simply, the "feel of place" (Degen 2008, 2010, 2014). The aim is to inscribe particular places within the city with a specific set of meanings and "ways of being, feeling and acting with the brand" that lead to the creation of particular urban lifestyles (Masuda and Bookman 2016, 171). In recent years this combination of urban redesign, social planning and branding has been defined in urban policy under the umbrella term of "placemaking".

Placemaking was a fundamental idea in the work of Jane Jacobs and William H. Whyte in the 1960s and 1970s to create more human-centred cities and as an oppositional stance to modernist urbanism's perceived placelessness, and has since underpinned the aims of urban design (Aravot 2002). Since the late 2000s the concept of placemaking has witnessed a resurgence and is regarded as a key ingredient, or even formula (see https://www.pps.org/article/what-is-placemaking), for successful urban regeneration to enhance a city's attractiveness to the creative classes (Lew 2017), particularly following the influential and highly contested work of Richard Florida (2005; for a critique see Peck 2005; Wilson and Keil 2008; Mould 2015 amongst many others). Thus, placemaking is far from a neutral concept but regarded critically as a key element of gentrification processes (see for example Wilson and Keil 2008; Montgomery 2016) as it tends to involve a physical re-design of places and a deliberate engineering of the social life of the neighbourhood. Placemaking has become an economic development tool: "[a] process of creating quality places where people want to live, work, play, shop, learn, and visit" (Wyckoff et al. 2015: vi quoted in Lew 2017, 453) where design professionals and governments follow a series steps and tools to promote infrastructure growth (Lew 2017).

During this process, the image of a neighbourhood is an important element to be managed and shaped as it informs perceptions and expectations of a place. As Zukin (1991) explains, since the 1960s with the development of new media technologies, the intensification of advertizing and the expansion of urban elites has led to an extensive critical infrastructure of media outlets from guidebooks to newspaper reviews, and now websites, bloggers, or Instagram posts amongst many other digital outputs that mediate ever more complex urban consumer spaces. These images and associated discourses about particular neighbourhoods are central in shaping the perceptions of those living or visiting a place and strongly shape the cultural recoding of places (Miles and Miles 2004) to create a "new urban imaginary":

> [A] coherent, historically based ensemble of representations drawn from the architecture and street plans of the city, the art produced by its residents, and the images and discourse on the city as seen, heard or

read in movies, on television, in magazines or other forms of [digital] mass media (Greenberg 2000, 228).

As Greenberg further explains, a number of urban imaginaries coexist and compete with each other in any locality. Moreover, as she expands by analysing the development of urban lifestyle magazines, the power and form of the precise mediation of urban imaginaries goes hand in hand with the development of the latest media technologies. If we consider the digital to be the most transformative and wide-reaching technological development since the start of the twenty-first century the question that arises is: how are digital technologies reconfiguring the planning and branding of neighbourhoods such as the Smithfield Market area? And, what kind of spatio-temporal relations do they convey?

Let us start by examining the planning aspects. As we have argued elsewhere (Degen 2018) planning in itself is a deeply temporal activity: "[A] continuous process [...] of choosing strategically through time" (Friend and Hickling 1997, 1; see also Abram and Weszkalnys 2011; Myers and Kitsuse 2000). And, one could argue, always future oriented. First, planning provides a tool and practice to manage the present, "of governing and organizing the relationship between the state, citizens, and other organizations whether private, commercial, or public" but, also it is "the transition over time from current states to desired ones" (Abram and Weszkalnys 2011, 3–4). Hence, much planning practice consists of preparing for future activities by trying to organize, predict and manage the spatial outcomes of future times. This is because urban space is highly unpredictable and "messy" and strategies, plans or future projections aid to "tame urban complexity" (Hoch 2009). Second, this future planning is informed by the past history of a place and present technologies, values and planning trends. Thirdly, as the future is the "not yet" (Adam 2006), it needs to be built as a performative trope into urban change. This is done through two main features: first, forecasts and projections of future economic growth, environmental sustainability, demographic change, estimated future visitor numbers; and second, through the construction of scenarios, visioning, and backcasting which aim to provide assurances to investors and engage the general public on an affective level with what the future will feel and look like. Thus, for future landscapes to be communicated effectively they need to be *visualized* convincingly. It is in this process where future urban imaginaries, branding and placemaking strategies start to merge.

Advances in digital technologies since the 1970s mean that computer generated images (CGIs) have become the common means for architects and developers to plan, visualize and market future urban developments. Digital visualizations created through the use of visualizing software applications such as Sketch Up, Rhino and Studio Max make it possible to compose carefully crafted images of buildings set within a future and imagined urban context as part of the design and planning process. Indeed, CGIs have now become the most common type of image media used to visualize and market future urban redevelopments and their envisaged social uses (Rose et al. 2016). They have become such a ubiquitous part of producing and marketing contemporary urban landscapes that we could claim that they are one way in which cities "are beckoned into existence by code" (Thrift and French 2002, 311).

The rapid and intense development of social media in the last decade has added another layer to how cities are experienced, mediated and imagined through digital technologies. For the purpose of this chapter we focus in particular on Youtube and Instagram. A range of scholars (Boy and Uitermark 2017; Rose 2016a, 2016b) have drawn attention to how social media and digital photography is changing our relationship with technology and space. It is important to highlight that one needs to differentiate between different uses of social media. Thus, organizations such as the Culture Mile or the Museum of London use social media as part of their communication strategy to promote events, activities and their brand. However, the ability by the general public to also be involved and take professional quality photographs, especially through the filters and editing facilities of Instagram, "challenge the distinction between professional and amateur and strategic and non-strategic" (Thelander and Cassinger 2017, 7). It is thus important to start examining what kind of power relations are being forged between these various urban digital imaginings, branding and spatial developments.

In the rest of the chapter we bring together these related areas of scholarship to outline a theoretical and empirical application that places the various temporalities of urban placemaking and digital visual imaginings in relation to what Sharma has described as the power-chronography which "provides a politicization of time that dispels individualistic accounts of time and allows the social and relational contours of power in its temporal forms to emerge" (Sharma 2014, 14). We look at the intersections between future placemaking and digital imaginings to understand who has the power to shape the urban imaginary, who is addressed by it and who is left out because "there is a politics and ethics of and to temporality and the future where futurity is actively involved in the making and remaking of difference and inequality" (Coleman and Tutton 2017, 444). Let us now turn to an overview of the development of the Culture Mile in relation to the relocation of the MoL.

The Culture Mile and the relocation of the Museum of London

The beginning of the Culture Mile (CM) in the City of London can be traced back to 2010 when the development of its first Cultural Strategy was approved. The Corporation of London, the City's municipal governing body, established "The Cultural Hub Working Party" which stated that its "vision for 2017 is to see the City's identity as a cultural hub strengthened in its own right, alongside its status as a financial centre" (Cultural Hub Working Party Report to Policy and Resources Committee 2013). It is important to highlight that the Corporation is the fourth largest cultural funder in the UK due to hosting in its borough many cultural institutions, yet is rarely acknowledged as such, and "[t]he hub would be both a *visual area* that invites people in *to experience its cultural offering* and a *collaborative hub* between renowned institutions [...] to draw in more visitors to this area and increase the exposure of, and enhance the quality of provision by, these renowned cultural institutions" (Cultural Hub Working Party Report to Policy and Resources Committee 2013, emphasis by authors). Publica, a private urban design practice, was commissioned by the City of London to develop a strategy for this cultural hub, as part of the Barbican and Golden Lane Area Enhancement Strategy, with the aim "to deliver a comprehensive identity for the area which will resonate and attract audiences from around London, the UK and the World" (Publica 2015). What stands out in this document is how Publica discusses the changing trends in culture within cities, suggesting that culture is increasingly provided in public space rather than merely within buildings and therefore there should be more collaboration between cultural institutions to create shared programmes of public realm events.

In 2016 the working group requested £100,000 from the City of London funds to employ marketing and communication experts to come up with a more distinctive brand for this cultural area. They employed Jane Wentworth Associates and Pentagram to complete a brand strategy which set out four clear values for the area: joined-up, generous, agile, experimental. They developed the name "Culture Mile" and an associated suite of imagery – logo, website, promo videos – and Culture Mile was officially launched in July 2017. It encompasses the mile from Farringdon to Moorgate and is a public-private partnership umbrella organization endorsed by the Corporation of London which includes the Barbican, the London Symphony Orchestra and the Museum of London. The City of London's 2018-2022 Cultural Strategy highlights the links between the urban redesign of the area and culture by stating that its first objective is to "[t]ransform the City's public realm and physical infrastructure, making it a more open, distinct, welcoming and

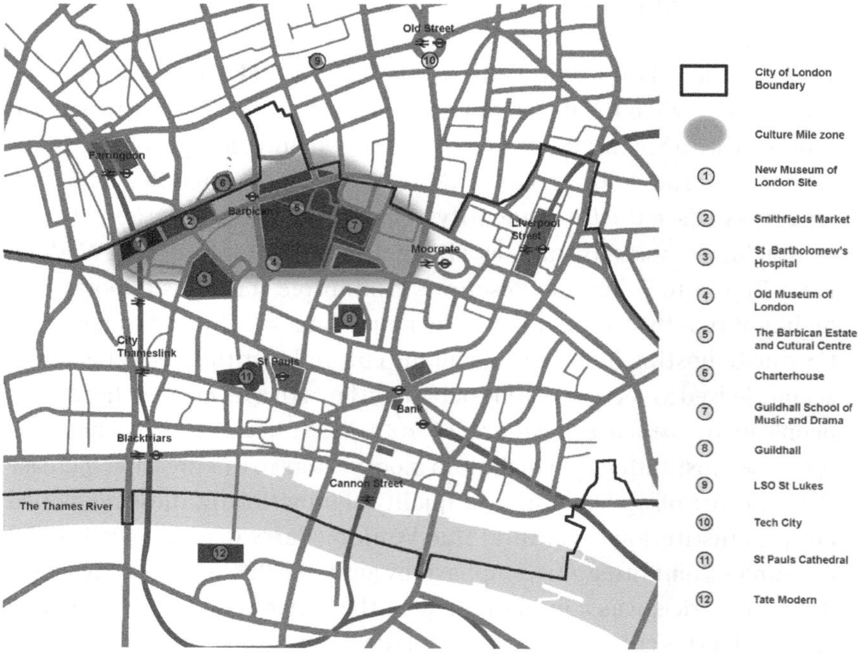

Figure 5.1. Map of area and Culture Mile. Courtesy of Isobel Ward.

culturally vibrant destination" (City of London 2018). This is implemented by the Corporation of London's planning department through their Look and Feel programme which is applying the recommendations made by Publica for their public realm enhancement programme. The second objective is to "[d]evelop Culture Mile in the north west of the City which will become an exciting destination for London and act as a catalyst for change across the rest of the Square Mile" (City of London 2018). Culture is clearly regarded here as a changemaker and there are clear links noted in the strategy between culture and commerce, suggesting that culture will become a new revenue for the City of London.

We can see here that the CM regeneration project allows an enactment of a particular "active" version of the future to contrast it with a "passive" present. Since the CM's launch, discourses in the media hint to "the potential", "genuine regeneration" (Pickford 2017) and "major destination" of the CM – that will "deliver new experiences for everyone" (Kenyon 2018). The MoL's relocation is being described as a landmark project that "will establish it with an international public" (Kenyon 2018) as cultural institutions are presented as an explicit part of the city's economic revitalization programme. Implicit in this future scenario building are the promises for a "better" material and temporal order, implying that the already present needs improvement: "Plans can be constructed to avoid undesirable futures, to make desired forecasts come true, or to create new, more desirable futures" (Myers and Kitsuse 2000, 223).

Turning now to the MoL specifically: the MoL has the largest urban history collection of the world (Kennedy 2016). It is a charitable institution funded by the City of London Corporation, the GLA and a range of benefactors. For its move to its new site in West Smithfield Market, by 2017 it had raised £110 million from the City of London Corporation, £70 million from the Greater London Authority and City Hall – their largest investment in a cultural initiative – and it needed to raise another £70 million from the private sector, individuals and charitable foundations (BBC 2017). Planning permission has been submitted in December 2020 with plans to start construction soon after.

In interviews with museum curators and managers, they explain that the need to move has been longstanding, due to the geographical location of the museum, housed above a busy roundabout near the Barbican, not easily accessible and discouraging to visitors: one needs to follow an array of escalators, steps and narrow corridors to be able to find it. Curators especially point to a desperate need for more space for its 7 million objects which are mainly kept in enormous warehouses in the East End. Thus, in their view the new venue in West Smithfield provides the museum an opportunity to reinvent itself and transform its relationship with the city and its publics. Hence, the MoL is particularly eager to develop a porous relationship between the Smithfield neighbourhood and its surrounding locality. As the Lead Curator of the New Museum states: "The vision for the new museum is that its whole look and feel will embody London. It will capture the essence and personality of the restless and creative city, including its past and present sensory experience" (Werner in Degen et al 2017).

The Museum of London's multiple imaginaries

Let us start analysing how the New Museum of London is visualizing itself in the future through the umbrella organization of the Culture Mile and through three kinds of spaces, each with different temporalities: a) urban space, b) strategic planning space, c) social media space. The data for this discussion was gathered from a six month pilot study collaboration with the MoL in 2017 which set out to examine the changing sensory identity of the Smithfield area from past, present and future. The multimethod study included ethnographic research, historical and contemporary planning research, interviews with key stakeholders, a vox pox survey with 110 members of the public as well as a sensory and temporal mapping of the area. The aim of the study was to gain insights into how the character of the area has evolved over the centuries, Smithfield's current identity and perceptions amongst the general public, and what the expectations of Smithfield's diverse publics were for the move of the New Museum of London (see www.sensorysmithfield.com). During this pilot research, we noticed an increased use of digital visualizations to promote the area and envisage its future and have since then been conducting further interviews, ongoing online research of social media and regular ethnographic observations of the area.

a) Urban space

To understand the placemaking at stake in this case study and how digital technologies are involved in this process, we start with situating the current senses of place as perceived by those using the Smithfield area. The concept of placemaking can be traced back to phenomenological cultural geography traditions aiming to understand how individuals or communities develop a sense of place over time and through their personal experiences and corporeal engagements with place (Relph 1976; Tuan 1977). This sense of place can be deeply personal or relate to more communal structures of feeling that have developed over generations to shape the attachments of particular groups to place (Pred 1986). As Lew further explains, "[t]hese are mostly organic, bottom up processes, whereby places are claimed and shaped through everyday, and often mundane social practices" (Lew 2017, 449). However, as he explains, increasingly this organic evolving of a sense of place is replaced through "placemaking" which refers to the strategic and planned work done by place branding organizations and the conscious positioning of particular media narratives, orchestration of events and stylization of the urban environment to construct specific place identities and meanings of place. Everyday engagements with places whether as a local or visitor are complex performances in which the combination of imagination, embodied experience and materiality of place create diverse forms of place consumption and engagement (Edensor 2001). When people engage with places, they draw on representations of place such as those in social media for example, to plan and inform their engagements, highlighting how the three spatialities we analyse are interconnected.

The main characteristic of the physical space which our respondents remarked upon was the juxtaposition of diverse sensory and temporal

Figure 5.2. A juxtaposition of architectural styles. Courtesy of Camilla Lewis.

experiences which creates a unique place identity in the Smithfield area. This stems from a temporal juxtaposition in the built environment which features buildings from a diversity of historical periods: the grand Victorian market building, surrounded by street layouts that have remained unaltered since 1870 and follow mainly a Medieval pattern. Next to it buildings date from various periods including Medieval monasteries, Victorian housing, the Barbican housing estate which was built in the 1960s to the east, and to the west a glittering façade of high glass buildings constructed in the 2000s.

The temporal juxtaposition within the built environment is intensified by a diversity of lived temporalities by various social groups whose distinct uses of the neighbourhood generate particular daily and weekly rhythms and overlapping, clashing sensescapes in the public spaces of Smithfield. The use of public space varies greatly across the times of the day as different demands shape and conflict in space: from the arrival of meat lorries that loudly take over roads from 11pm creating traffic jams; market workers in their white robes shouting orders and young clubbers jostling and sharing pavements between 11pm and 7am; a diverse fleet of market customers from exclusive restaurants to halal butchers arriving at the market between 2am and 6am; and a mixture of city and creative workers from 8am to 6pm mingling on streets with builders, hospital staff, couriers, tourists,

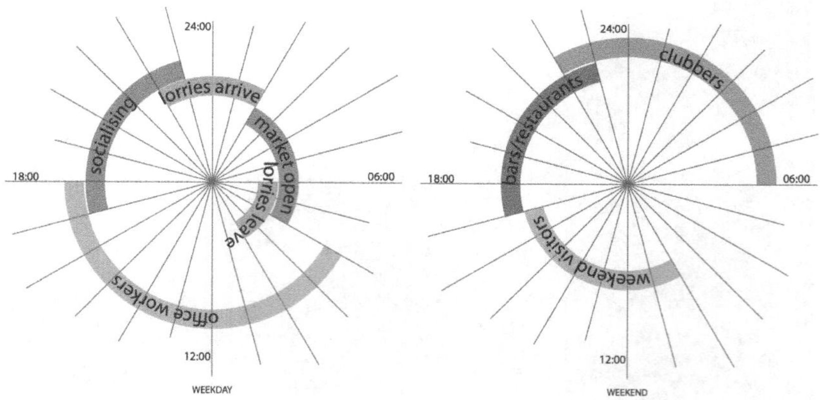

Figure 5.3. Temporal rhythms of Smithfield market area. Courtesy of Isobel Ward.

and restaurant visitors at nights and weekends. One of our interviewees described the idiosyncratic feel of the area as follows:

> It's got a bit of bite to it [...] it's a bit grit, a bit of edge, a bit trendy here and there [...] This place is alive at 3am in the morning [...] You've got weird little cocktail bars around the corner underground. There's the market just around the corner. It's all this mash of [...] it just feels it's got that little edge to it, which is interesting (vox pop June 2017).

All of our vox pop survey respondents commented on how it is precisely the variety of social uses that give the area a particular vibrant character. One local resident we interviewed described Smithfield as "buzzy" and commented on how unexpected encounters were commonplace, as different groups found themselves "cheek by jowl" in this area of the city.

These results show that it is precisely the entanglements of past and present buildings and traditions in the landscape and uses of place that give the Smithfield area its unique sense of place. In addition, the temporalities of the past are woven through this environment not just with the continuity of spatial patterns of the built environment or the concentration of historical sites, but also through an imagined narrative of a place which has contested wider processes of change. This is an argument that has been taken up by conservationists and in the media, and is continuously presented through the visual imagery displayed by hotels, pubs and heritage information boards on the market in the area. For example, one of the hotels has a series of historical images and Dickensian quotes displayed on its windows to highlight the historical linkages and continuity in the look and feel of the area, highlighting that the Great Fire was extinguished just before it reached the Smithfield site, and bomb damage to the buildings during the Second World War was not extensive. Fundamental changes which have taken place in its social, spatial and sensory history are not mentioned in these narratives such as for example spatial restructuring by the Corporation of London that can be dated back to the twelfth century, Victorian attempts at urban sanitation, changes in the social fabric of the neighbourhood or changes and developments in transportation facilities such as the building of the rail and underground routes that cut through the area, to mention a few. A strong theme that emerged in our interviews is that Smithfield is generally described as having managed to "buck the trend" of the high speed capitalist city and is therefore perceived as an area stagnant in time.

This perceived narrative of continuity has been mobilized in studies intended to imagine or plan future regeneration projects. While there have been several endeavours since the 1970s to regenerate this area, future developments have always been halted. Two attempts to demolish West Smithfield were stopped after public inquiries in 2005 and 2014 prompted by local heritage campaigners who argued that the market buildings made "a significant contribution" to the character and appearance of the area. And, a report prepared for English Heritage, argued "the fragile identity – defined by its architectural character, streets, places and activity patterns" should be "respected and reinforced rather than ignored" (Farrell 2007, 3). Hence, when initial planning strategy meetings took place to discuss the current plans for redevelopment, the heritage status of the site and market traders' concerns had to be taken on board. Local authority planners are aware of this and there is a constant tension between maintaining historic sites and redeveloping for a changing city – thus the MoL's plan to move in has been regarded by many market traders and the general public as positive, able to negotiate these concerns and be an enabler for the current regeneration of the market.

These concerns have strongly influenced the creation of the branding campaigns by the New Museum of London of its new location which is advertised as the "Museum of Londoners" on huge billboards around Smithfield's current spaces which visually frames portraits of individuals engaged in their particular working practices within the wider history of the area. The aim was to create: "[A] campaign that is the museum of the people who live and work here. It's the Museum of Londoners. So, we focus around the people who are working in this area now, either in the markets or in the cafés or in the hairdresser training place [...] so, we're very kind of

Figure 5.4. Museum of Londoners and Crossrail advertising. Courtesy of Monica Degen.

consciously, deliberately wanting to affect the neighbourhood" (Director of Transformation MoL). The images include a market butcher, Crossrail engineer, trainee hairdresser and local café owner amongst others. The portraits stand out for their crisp colours, boldness and intimate feel and reflect London's cultural and ethnic diversity. The hoardings cover the buildings to be developed and provide a visual reference point and landmark as they melt with the surrounding urban landscape, providing a scene that individuals then capture on their mobile phones. They were taken by 20 year old local photographer Vicky Grout who has developed her career by documenting the lifestyle and development of London's grime scene by first posting her pictures on Instagram. These billboards are physically dominating the built environment earmarked for transformation, and physically announce an impending sense of change, yet are clearly based on local links.

They stand in contrast with a second type of image that are advertising the forthcoming urban change on hoardings: namely Crossrail CGI images and those by developers advertising new residential spaces. These billboards project very "run of the mill" CGIs: featuring soft colours, a corporate, cleansed, and smooth vision of the future – erasing any link to the existing physical location and alluding to the creation of an imaginary local community of future commuters, residents and visitors – and very much ignored by passers by.

b) Strategic planning space

The second spatial dimension we would like to analyse is the strategic planning of the New Museum of London within the Culture Mile which explicitly draws on digital visualizations to present a vision of the future and prepare the public for the imminent spatial changes. A series of exhibitions to inform and consult the general public took place in 2019 at the West Smithfield site and within the MoL heavily featured CGIs designed by Stanton Williams and Asif Khan, with conservation architect Julian Harrap. The CGIs feature heavily on the current MoL website (https://museum.london/). It is important to note that planning permission still needs to be approved at the writing of this chapter and that the architects and museum curators are in the first stage of the development, i.e. producing a brief for the planning application.

A vital issue for the New Museum is enhancing its visibility and permeability – referring to the Museum's links with its surroundings spatially,

Figure 5.5. (a) Proposed Campus. Source: Stanton Williams and Asif Khan, © Secchi Smith. (b) Proposed West Smithfield at Night. Source: Stanton Williams and Asif Khan, © MIR.

culturally and socially. The move to West Smithfield is expected to increase visitors from 750,000 to two million a year. According to the Director of Transformation, the problem with the MoL's current location is that it was based on "a concept of a cloistered, quiet space in the heart of the City. Very inward-looking, which doesn't chime with twenty-first century agendas for public museums". Hence, not surprisingly the CGIs of the future development depict activity: moving people, and those that stand still are clearly taking pictures or actively observing the new environment. The CGIs' lighting also enhances feelings of openness and permeability of the buildings as sun rays shine through and images are taken from a pedestrian's point of view, clearly highlighting entry points and featuring people moving across different spaces. As one of the architects involved in the redesign of the building explains, the new Museum of London will integrate spatially and conceptually with the city:

> The city will be drawn into the Museum and will have many different partners working inside who they will engage with. It's like the idea of the public house. We are reinventing the public realm. [...] The entrance to the market is vital. We want there to be lots of entrances, maybe about seven. This is going to be a democratic realm where people can pass through, meet up or have a coffee.

While many of the curators we spoke to were concerned about the curatorial content and narratives of the New Museum, the management were concerned about how to create a landmark venue which attracts both local and global audiences.

Let us analyse this in more detail. The market building is part of a heritage site which allows the new museum to enter into a temporal dialogue with its surroundings while simultaneously providing its differential quality, its landmark value effect. The historical market is crucial to the New Museum's conception, however the existing old, abandoned and weathered buildings are reinterpreted in the CGIs, lifted into the future, and appear cleansed, light and mixing elegantly with contemporary design features. The CGIs are central in conveying this experiential aspect of the planned museum. While drawings of planned buildings always attempt to make them attractive, CGIs can produce particularly atmospheric and photo-realistic views of buildings, often glowing at dusk and thus tapping into the affectual and kinaesthetic tactility of the city, thus taking the viewer into the future feel of place (Degen et al. 2017).

While it is acknowledged that a redesign of the existing buildings is part of the broader rebranding of the area, there is also an attempt "to hold onto the character of the area" (architect) and to "reflect the melting pot character of the area" (architect). Yet, an inherent tension for the design and planning of the building is between what is "for the best interest of the museum for the next generation" and "the [existing market] building telling us things. So, the building will say, 'I'm not supposed to do this', or 'I'll let you do this' to me" (Director of Transformation MoL). Thus, the existing market building might not allow some of the contemporary design features suggested which leads architects and planners to find compromises. However, this is not explicitly acknowledged in the CGIs which not only present a "finished design" but invoke clearly the future uses and publics that will access these buildings. This illustrates how "[t]he future is not a disconnected end-state that exists only in the future; instead the future should be viewed as a continuous unfolding of time that is rooted in past and present" (Myers and Kitsuse 2000, 225). The present and future uses are incorporated in this envisaged building through the CGI: "[T]he move is a way of changing the museum, doing something different and reinventing it for a twenty-first century audience" (Director of Transformation). This change, in his view, will be driven by the architectural experience that the New Museum provides: "It's going be the sort of uniqueness of the buildings and kind of experience of being in the buildings that is going drive people [to visit]." Hence, the brief for the architectural competition emphasized strongly the provision of different uses of the buildings:

> The ambition that the museum is sort of integrated into the public realm [and] that it becomes something that is a bit more like Southbank Centre, in that people feel comfortable using these buildings for their own purposes at different times of the day or night. And it's not necessarily the fact that people are going to be driven by a kind of cultural imperative that they're coming to see something about history, or coming to see something about a particular London narrative or an object. But [rather], that the building has a greater sense of ownership by Londoners. [...] that translates into, for example: how could Londoners be part of, and leave a part of them in, the museum? (Director of Transformation)

So, who are these Londoners? The analysed CGIs suggest The New Museum of London is planned to be a twenty-four hour museum, reflecting the current twenty-four hour character of the neighbourhood; but appealing to different publics than those currently creating these twenty-four hour rhythms. So, a move away from the current clashing melange of meat market workers, clubbers, hospital workers and city workers to a new consuming public consisting of families, tourists and a diversity of young people. As mentioned above, the New Museum's ambition is to attract visitors and it promises novel experiences including a twenty-four hour bar area, boutique retailers, and an array of events such as the London Fashion show to entertain new publics. One could argue that these competing discourses and imageries reflect the various and sometimes competing roles that contemporary museums have in contemporary society from curators of knowledge to entertainment venues.

We have shown in this section how the strategic planning imaginaries draw upon overarching spatio-temporal characteristics currently present in Smithfield such as the re-use of historical buildings or the twenty-four hour uses of place. Yet, while these spatio-temporal characteristics of the built environment's current uses have emerged over time from organic processes and are produced by a variety of social groups, we can see that the combination of planning strategies and digital visualizations aim to produce a smooth new choreography of uses to attract specific publics. The digital visualizations portray a spatial re-organization of place, new activities taking place in the public spaces and a clear transformation of the current sense of place.

c) Social media space

In this last section we want to reflect on the role that social media space plays in the placemaking of the MoL's imminent move. Social media branding has become part of communication strategies of organizations, and neighbourhood redesign schemes are no exception. Society's move to the digital presents a challenge for cultural institutions, as a quotation by the Chief Digital Officer of the New York Met used at a presentation explaining the concept behind the Culture Mile illustrates: "People ask me: is your biggest competition the MoMa? No, our competition is Netflix. Candy Crush. It's life in 2016" (Wentworth 2018).

Social media activity in regards to the new MoL has taken place via a number of channels. The MoL's own digital communication about their new location has been gradually building up since 2016 as discussions with different stakeholders involved or affected by the move have evolved. Hence, the MoL has been digitally branding its move with a series of YouTube videos which started in 2016 and more recently via its designated website (https://museum.london). It has branded itself to a much lesser extent via Instagram, mainly drawing attention to its YouTube videos. The MoL's initial promotion of its new location via the YouTube videos celebrates and presents the current feel of the Smithfield Market area, presenting its history through the built environment, its diverse social uses, temporal rhythms and sensescapes which characterize the identity of the area, using mostly filmed footage and photographs. The videos do not show any of the CGIs of the new location. Smithfield's current physical and social uses are clearly presented here as a backdrop for the New Museum as stated on their website:

> We believe London is the world's greatest city and we are uniquely placed to tell its story, but only if we have a showcase worthy enough. So, our ambition is to do this at the heart of a new cultural hub in the City of London with outstanding links to the rest of London and the world, and in doing so we will become one of London's top five most visited museums. (MoL website)

From a branding point of view one could argue that the contribution that the current market and its everyday practices provide to the temporal and sensory set up of the area are kept as an important feature of these films and provide the area's unique selling point.

A much more intensive and public-facing branding campaign for the cultural regeneration of the area has been the setting up of the Culture Mile since late 2017 which allowed the Corporation of London to bring together four of its key cultural institutions, including the Museum of London. Because of the lack of a masterplan for the entire Smithfield area due to complex boundary issues with different councils, the network provides a cohesive identity and performs a united vision. The loose network, whose staff are dispersed across the various cultural institutions involved and the planning department of the Corporation, helps to imagine a coherent whole and supports powerful actors to perform and own a particular vision of the future of this area, the Culture Mile, in order to:

> [T]ransform the area, improving their offer to audiences with imaginative collaborations, outdoor programming and events seven days a week. Links between venues will be improved and major enhancements to the streets and wider public realm will enliven the area which, as Culture Mile expands and flourishes, will be regenerated (see Corporation of London website).

A crucial focus for the Culture Mile has been the "activation" and animation of public life in streets in the neighbourhood, as the area is planned to become more residential and shift its reputation of a "dead" city centre adding to the creation of a new place imaginary. The public envisaged to be attracted to its streets includes "city businesses, office workers, residents, schools, cultural and commercial occupiers, creative industries partners and potential investors and funders" (www.culturemile.london). Alongside traditional media such as posters in London Underground stations and Time Out London, social media, especially Instagram, has been one of the main media through which the Culture Mile events have been marketed and branded with the view to bring particular publics to the area. From November 2017 to May 2020 Culture Mile Instagram features 125 posts and has 2305 followers. Thus for example the "Culture Mile Nights" event was explicitly branded to under forty-year olds:

> Our Instagram like tessellation of images, by curating that and making it look really appealing for a certain audience, was really key for marketing Culture Mile Nights [...] That was like a really kind of a pivotal moment in terms of how we use that particular channel. Because we knew that Instagram [is] attracting an audience of between eighteen and thirty-five, late thirties, for this particular event they were like the key audience for it. (Marketing Manager CM)

Indeed, when programming events the creation of experiences, or "memorable moments" (Pine and Gilmore 1999), captured and shared via social media, especially Instagram, are an important feature to be taken into account. The capturing of experiences and their distribution via social media influence the placemaking strategies of the Culture Mile showing how the planning of activities is linked to a conscious awareness of how they will be distributed and shared:

> If you think about what people use to curate their own experience of an area, that's one of the key things [we consider]. Not just in terms of like constructing outwardly that experience. But for themselves. It's not just about showing to other people that you've been to a place. It's also, for you, like how do you remember the place? What were the things that you were really excited by? So, [social media] has absolutely changed placemaking. [...] It certainly is something that we consider when we're looking at public programming: how is this going to be remembered? How is this going to be like distributed or shared? It's not the number one guiding principle, but it's absolutely something that's in the back of my mind. (Marketing Manager CM)

This supports Tiidenberg's suggestion that three key features of the social media visuality are "(1) networked visuality which centralizes sharing, (2) emplaced visuality which centralizes movement and location; and (3) conversational visuality which centralizes personal interactions via visuals" (Tiidenberg 2018, 14). Similarly, these three aspects are central in the way the redevelopment and transformation of the Smithfield area is mediated and branded via Culture Mile posts: images are shared across different publics; particular views, places and moments are emplaced visually and temporally; and the personal experience of a sense of place is further communicated through remarks, hashtags, likes and reposts. We can understand posting of place tags as conscious markers of a curated place identity by organizations or users and as a symbolic claim to place (Budge 2020) which are shared with others.

While the focus of this chapter is on how social media is used by organizations in placemaking strategies, some of the insights in studies on amateur use of Instagram are relevant for this analysis. For example, the work of Boy and Uitermark (2017) on how individuals portray different neighbourhoods on Instagram, shows how both the production and the content of social media can reiterate very particular understandings of places and people as only specific places or scenes are depicted and shown in particular ways. Precisely because screens, including phone screens, are part of the everyday staging of urban life, their contents appear and intervene in the experiencing of cities. In their study of how different neighbourhoods in Amsterdam appear on Instagram, Boy and Uitermark remark that:

> Instagram users selectively and creatively reassemble the city as they mobilize specific places in the city as stages or props in their posts. Instagram images, in turn, become operative in changing the city [...] users view the posts of others. (2017, 613)

This picturing and its circulation serves to amplify, in their case study, the gentrification of those neighbourhoods. Similarly, one could argue that the postings by the Culture Mile serve to guide people to explore particular places and events in the Smithfield area but not others, presenting a curated and partial account of the neighbourhood. These social media posts strongly influence the visual imagining of how place is perceived, experienced and understood. In the case of the Culture Mile, its posts relate to their organized events, public art projects and the cultural events promoted by the cultural institutions that are part of it. No images appear of the working market or other current everyday uses of the neighbourhood.

In another digital campaign in 2018 the Culture Mile sponsored a series of films by poets called "Between the Storeys" which were distributed via YouTube and other social media. Over 6 months three poets and a film-maker engaged with the neighbourhood and its inhabitants and workers producing specially commissioned poems "to animate the stories, histories and experiences of different communities living within the Smithfield area" (https://www.culturemile.london/betweenthestoreys/betweenthestoreys). These poems and films evoke and reflect the past and present culture in the Smithfield area, both in terms of its rich and varied built environment and mixed social uses. And, as the market area has not been regenerated yet, the poems reflect the life and the sensations that can be encountered in the present moment. All three poems combine particularly sensory evocations of past, present and future, paying particular attention to the clashes of social groups and sensations: "A Community of Souls" reflects on community and regeneration, young and old, new buildings and historic buildings; "Underlines/Overheard" evokes the current and past railways and uses; and "Began in Fabric" draws out the similarities between the comings and goings of the night club Fabric and the meat market.

These urban imaginaries overlap in many ways with the perceptions of the urban landscape by people living, visiting or working in the area; the history ingrained in the built environment, the activities in and around the meat market and the idiosyncratic feel of place captures these artists' imaginations and artistic outputs. The future strategic planning space is not mentioned or, if mentioned, evoked as a threat to the delicate "street ballet" (Jacobs 1961) that characterizes this area. These online videos largely simultaneously represent the area based on the present sensations, everyday rhythms and communities one can encounter but that will gradually be stripped away in the future through the cultural regeneration and inevitable gentrification of the area. As the videos are commissioned and featured by the Culture Mile we can argue that they brand the "present feel of place" by drawing on the existing urban space, current lifestyles, atmospheres and senses of place as a unique selling point even though the forthcoming cultural regeneration will inevitably change this.

Conclusions

This chapter has examined the cultural regeneration of the Smithfield Market area by analysing how future urban imaginaries are mediated and visualized through a range of digital technologies and across a range of spatialities to advance urban development. This is used as a tool by planning and building professionals to create support for the planning application, sell the development to future investors and to prepare the public for the new look and feel of a neighbourhood. Urban design, branding and placemaking activities clearly merge in contemporary regeneration processes. Indeed, this case study has shown how digital technologies such as the production of CGIs and social media branding can be viewed as digital foundations to prepare the ground, showcasing and promoting future urban imaginaries *before* a redevelopment takes place: visibility is key in order to compete on the global catwalk (Degen 2008). The digital visualizations make the future a perceptible, present vision, yet only represent selective narratives and characteristics – in this case a future imagined in line with the agenda of creating a cultural quarter to attract new audiences to the Smithfield area.

We have shown how a diversity of future urban imaginaries are convened across different spatial representations in the development of the Culture Mile and the re-location of the MoL. We started by examining the perception of the urban space of the Smithfield area by current users and argued that the MoL's Museum of Londoners hoarding campaign presented a future embedded physically within the urban textures and within the present daily life of the neighbourhood, showcasing existing practices of the neighbourhood and addressing local people. We then analysed the strategic planning space where the MoL presents a future aimed at a new global public through the use of CGIs which represent the future experiential aspects of the new development, providing suggestive and photo realistic imaginaries of the future uses. Lastly, we examined the branding of the transformation of the area and the new MoL location through social media campaigns via the Culture Mile. Here the area and its future are very much located in a current sense of place, drawing on particular aspects of its history, and embedded in the current urban fabric and everyday practices. Yet, in our view, branding the area selectively through art and culture also creates a particular visual aesthetic framing of Smithfield and its surrounding area as a consumable product – part of an urban lifestyle one can partake in when visiting the area. The branding uses social media to target simultaneously a global and niche public, interested in the area and specific historical and cultural events. We have highlighted how these urban imaginaries at times overlap, at other times differ from or, indeed, might change the current socio-spatial set up of Smithfield by drawing on the physical aspects of Smithfield Market, yet with a tendency to reframe and recode the social uses and cultural meanings of the area.

The power chronography, or politics of time, in the city come to the fore when we analyse who has the power to shape these urban imaginaries. Most people living, visiting and working in Smithfield say the meat market and its practices and cultures are part and parcel of what produces its unique sense of place, and want to protect and retain it. Those involved in planning and branding the future Culture Mile believe that the public desires different types of spaces and experiences. Yet, the exact demographics of these publics are not clearly defined for cultural organizations such as MoL as there is a tension between its commercial aims to become a "destination" and compete with other museums within London and globally, and its role as a civic, cultural and knowledge institution. The distinctive urban imaginaries visualized for these differing aims have one thing in common: they conjure the future by appealing to our sensory, affectual and experiential sensibilities. This is done through various means such as identifying with people's everyday practices in the present urban space; suggesting future uses and lifestyles through glamorous CGIs; via Instagram posts which disseminate selected intensified moments that showcase cultural events promoted through the Culture Mile; or showcasing evocative poetic films based on the current senses of place. The research has shown that the future relocation of the MoL within the Culture Mile is perceived, imagined and constructed through many forms and imaginaries illustrating that the future is far from being one dimensional but envisaged as multiple possibilities.

While the temporal logic of the capitalist city is linked to financial capital and movement, it is clear from this case study that the stagnation of urban redevelopment over years has allowed for different temporalities to emerge in the Smithfield Market area. This unique identity is cherished by the current publics working, visiting or living in the area who have adamantly expressed that they would not like the area to become another sanitized retail and service environment such as Spitalfields or Covent Garden. As Lew (2017, 462) highlights, approaching place making as both organic and planned "informs us of what we cannot control, and what needs to be given freedom to evolve in its own way and its own time". This is a challenging task and points to the paradox inherent in this project: the move of the Museum of London to Smithfield Market will irrevocably transform the area. While the museum is eager to land softly and become part of an existing neighbourhood, its own development and success will threaten and transform the uses and senses of place of this diverse area. Yet, its move also opens up an opportunity for the New Museum of London to re-evaluate, explore and re-imagine, maybe radically, what being a city museum consists of, its role in its locality and wider urban space.

Postscript: The planning permission for the New Museum of London was accepted on the June 23 2020 (https://museum.london). At the time of the writing of this article it is difficult to predict how the global Covid-19 pandemic will affect this project's future.

References

Abram, Simone, and Gisa Weszkalnys. 2011. "Introduction: Anthropologies of Planning — Temporality, Imagination, and ethnography." *Focaal* 61: 3–18.

Adam, Barbara. 2006. "Futurescapes: Challenge for Social and Management Science." Paper presented at the 5th Palermo International Conference on Social Time Retroscapes and Futurescapes, Palazzo d' Aumale, Terrasini, June 2006. https://citeseerx.ist.psu.edu/viewdoc/download?doi=10.1.1.101.1105&rep=rep1&type=pdf

Aravot, Iris. 2002. "Back to Phenomenological Placemaking." *Journal of Urban Design* 7 (2): 201–212.

BBC. 2017. "Museum of London Gets £180 Million Funding Boost." BBC News, January 24, 2017, https://www.bbc.com/news/uk-england-london-38732919?intlink_from_url=&/.

Boy, John D., and Justus Uitermark. 2017. "Reassembling the City Through Instagram." *Transactions of the Institute of British Geographers* 42 (4): 612–624.

Braun, Erik, Mihalis Kavaratzis, and Sebastian Zenker. 2013. "My City – My Brand: The Different Roles of Residents in Place Branding." *Journal of Place Management and Development*, 6 (1): 18–28.

Budge, Kylie. 2020. "Visually Imagining Place: Museum Visitors, Instagram, and the City." *Journal of Urban Technology*. 27 (2): 61–79.

City of London. 2018. "Cultural Strategy 2018-2022." City of London Corporation. https://www.cityoflondon.gov.uk/assets/Things-to-do/city-of-london-cultural-strategy-2018-22.pdf/.

Coleman, Rebecca, and Richard Tutton. 2017. "Introduction to Special Issue of Sociological Review on 'Futures in Question: Theories, Methods, Practices'." *The Sociological Review* 65 (3): 440–447.

Cronin, Anne M., and Kevin Hetherington, eds. 2008. *Consuming the Entrepreneurial City: Image, Memory, Spectacle.* London and New York: Routledge.

Cultural Hub Working Party. 2013. "Cultural Hub Working Party Report to Policy and Resources Committee." City of London Corporation. https://democracy.cityoflondon.gov.uk/ieIssueDetails.aspx?IId=35627&Opt=3/.

Culture Mile. 2020a. "Between the Storeys." Culture Mile. https://www.culturemile.london/betweenthestoreys/betweenthestoreys/.

Culture Mile. 2020b. Culture Mile. https://www.culturemile.london/.

Degen, Monica Montserrat. 2008. *Sensing Cities: Regenerating Public Life in Barcelona and Manchester.* London: Routledge.

Degen, Monica Montserrat. 2010. "Consuming Urban Rhythms: Let's Ravalejar." In *Geographies of Rhythm: Nature, Place, Mobilities and Bodies*, edited by Tim Edensor, 21–32, Aldershot: Ashgate.

Degen, Monica Montserrat. 2014. "The Everyday City of the Senses." In *Cities and Social Change: Encounters with Contemporary Urbanism*, edited by Eugene McCann, and Ronan Paddison, 92–111, London: SAGE.

Degen, Monica Montserrat. 2018. "Timescapes of Urban Change: The Temporalities of Regenerated Streets." *The Sociological Review* 66 (5): 1074–1092.

Degen, Monica, Clare Melhuish, and Gillian Rose. 2017. "Producing Place Atmospheres Digitally: Architecture, Digital Visualisation Practices and the Experience Economy." *Journal of Consumer Culture* 17 (1): 3–24.

Della Lucia, Maria, and Mariapina Trunfio. 2018. "The Role of the Private Actor in Cultural Regeneration: Hybridizing Cultural Heritage with Creativity in the City." *Cities* 82: 35–44.

Edensor, Tim. 2001. "Performing Tourism, Staging Tourism: (Re)producing Tourist Space and Practice." *Tourist Studies* 1 (1): 59–81.

Farrell, Terry. 2007. *Smithfield-Farringdon: Reinforcing Urban Identity & Character.* London: English Heritage.

Florida, Richard. 2005. *Cities and the Creative Class.* New York: Routledge.

Friend, John and Allen Hickling. 1997. *Planning Under Pressure: A Strategic Choice Approach.* 2nd edition. London: Butterworth-Heinemann.

Greenberg, Miriam. 2000. "Branding Cities: A Social History of the Urban Lifestyle Magazine." *Urban Affairs Review* 36 (2): 228–263.

Harvey, David. 1989. "From Managerialism To Entrepreneurialism: The Transformation In Urban Governance in Late Capitalism." *Geografiska Annaler: Series B, Human Geography* 71 (1): 3–17.

Hoch, Charles. 2000. "Planning Craft: How Planners Compose Plans." *Planning Theory* 8 (3): 219–241.

Hubbard, Phil, and Tim Hall. 1998. "The Entrepreneurial City and the 'New Urban Politics'." In *The Entrepreneurial City: Geographies of Politics, Regime and Representation*, edited by Tim Hall and Phil Hubbard, 1–23, Chicester: Wiley.

Jacobs, Jane. 1961. *The Death and Life of Great American Cities.* Harmondsworth: Pelican.

Kennedy, Maev. 2016. "Off to the Market: Museum of London Shows Off its New Smithfield Site." The Guardian, May 20, 2016.

Kenyon, Nicholas. 2018. "The Culture Mile that Will Transform Arts in the City." *Evening Standard*, March 16 2018. https://www.standard.co.uk/comment/comment/the-culture-mile-that-will-transform-arts-in-the-city-a3792026.html/.

Klingmann, Anna. 2007. *Brandscapes: Architecture in the Experience Economy.* Cambridge, MA: MIT Press.

Lew, Alan A. "Tourism Planning and Place Making: Place-Making or Placemaking?" *Tourism Geographies* 19 (3): 448–466.

Lucarelli, Andrea. "Place Branding as Urban Policy: The (Im)political Place Branding." *Cities* 80: 12–21.

Masuda, Jeffrey R., and Sonia Bookman. 2018. "Neighbourhood Branding and the Right To The City." *Progress in Human Geography* 42 (2): 165–182.

Miles, Malcolm, and Steven Miles. 2004. *Consuming Cities.* London: Palgrave MacmIllan.

Museum of London (MoL). 2021. Museum of London. https://www.museumoflondon.org.uk/museum-london/.

Myers, Dowell, and Alicia Kitsuse. 2000. "Constructing the Future in Planning: A Survey of Theories and Tools." *Journal of Planning Education and Research* 19 (3): 221–231.

Pickford, James. 2017. "Nicholas Kenyon on London's New 'Culture Mile'." Financial Times, July 20, 2017. https://www.ft.com/content/ab89e83e-6ca3-11e7-b9c7-15af748b60d0/.

Pine, Joseph B., and James H. Gilmore. 1999. *The Experience Economy: Work Is Theatre and Every Business a Stage.* Boston: Harvard Business School.

Pred, Allan. 1986. *Place, Practice and Structure: Social and Spatial Transformation in Southern Sweden: 1750–1850.* Cambridge: Polity Press.

Publica. 2015. "Barbican and Golden Lane Area Strategy." August 23, 2015. https://issuu.com/publicaassociates/docs/150826_181206_barbican_strategy_final/.

Relph, E. C. 1976. *Place and Placelessness.* London: Pion Ltd.

Rius Ulldemolins, Joaqium. 2014. "Culture and Authenticity in Urban Regeneration Processes: Place Branding in Central Barcelona." *Urban Studies* 51 (14): 3026–3045.

Rose, Gillian. 2016a. "Cultural Geography Going Viral." *Social & Cultural Geography* 17 (6): 763–767.

Rose, Gillian. 2016b. "Rethinking the Geographies of Cultural 'Objects' Through Digital Technologies: Interface, Network and Friction." *Progress in Human Geography* 40 (3): 334–351.

Rose, Gillian, Monica Degen, and Clare Melhuish. 2016. "Looking at Digital Visualizations of Urban Redevelopment Projects: Dimming the Scintillating Glow Of Unwork." In *Cities Interrupted: Visual Culture and Urban Space,* edited by Shirley Jordan and Christoph Lindner, 105–120, London: Bloomsbury.

Sharma, Sarah. 2014. *In the Meantime: Temporality and Cultural Politics.* Durham, NC: Duke University Press.

Thelander, Åsa, and Cecilia Cassinger. 2017. "Brand New Images? Implications of Instagram Photography for Place Branding." *Media and Communication* 5 (4): 6–14.

Thrift, Nigel, and Shaun French. 2002. "The Automatic Production of Space." *Transactions of the Institute of British Geographers* 27 (3): 309–335.
Tiidenberg, Katrin. 2018. *Selfies: Why We Love (and Hate) Them*. United Kingdom: Emerald Publishing Limited.
Tuan, Yi-Fu. 1977. *Space and Place: The Perspective of Experience*. Minneapolis: University of Minnesota Press.
Wentworth, Jane. 2018. "Culture Mile: The Creative Heart of the Square Mile." Presentation at Jane Wentworth Associates, November 11, 2018.
Werner, Alex. 2017. "Foreword: The Changing Feel of Smithfield." In *The Changing Feel of Smithfield: Exploring Sensory Identities and Temporal Flows*, by Monica Degen, Camilla Lewis, Astrid Swenson, and Isobel Ward. London: Brunel University.
Zukin, Sharon. 1991. *Landscapes of Power: From Detroit to Disney World*. London: University of California Press.

About the authors

Monica Degen is a Reader in Cultural Sociology at Brunel University London. Her research focuses on the politics of space with a particular interest in the ways sensory, temporal and emotional dimensions underpin urban culture and politics. In 2016 she was awarded a British Academy Fellowship to research „Timescapes of Urban Change". More recently she has been working on developing creative methodologies and digital tools to capture the sense of place of cities and the ways in which urban environments are stratified by power relations; see www.sensorysmithfield.com and www.sensorycities.com. Her forthcoming monograph, *A New Urban Aesthetic: Experiencing Urban Change Digitally* (with Gillian Rose), explores how digital visualizations are transforming urban experiences.

Isobel Ward is an AHRC-sponsored PhD candidate in the Department of Geography at King's College London. She has carried out research in London amongst communities that are undergoing considerable urban renewal around issues of mobility and migration, place attachment and social ties to look at processes of un-making and re-making home. She has an interest in sensory, visual and digital methodologies to explore emerging urban spatialities.

6. Animated Embodiment: Seeing Bodies in Digitally-mediated Cities

Gillian Rose

Abstract
This chapter explores one aspect of how embodiment is reconfigured in digitally-mediated cities. Urban scholars have paid considerable attention to the immense streams of digital personal, environmental and geospatial data flowing through and beyond cities, but less to the data that becomes images of various kinds. Yet at the interface between digital images and their viewers, particular kinds of bodies coagulate: flesh is organized visually and spatially. The chapter argues that long-established ways of seeing and constituting bodies in cities are being joined by other ways of seeing other forms of embodiment, which are caught up with the specific dynamics of urban data circulation. The chapter argues that animatic embodiment is becoming part of the digital refiguration of urban space.

Keywords: mediation, animatic spatiality, embodiment, social difference

Introduction: mediating posthuman embodiment

To be human – more specifically, to be constituted as more, or less, fully human in the cultures and empires with which this chapter is concerned – has always been deeply imbricated in being visualized in particular ways (Edwards 1992; Rettberg 2014). Many but by no means all humans have mediated both their own humanity and that of others through various technologies that visualize fleshy corporeality in particular ways. Historically, many images have emphasized face, skin and hair, and the clothing and objects which adorn them, from grand oil paintings to documentary photography to police mugshots to the family snap to the selfie. The visual devices which picture bodies in these ways have focussed on surfaces: skin,

Rose, G. (ed.), *Seeing the City Digitally. Processing Urban Space and Time.*
Taylor & Francis Group, 2022
DOI 10.5117/9789463727037_CH06

hair, textiles, jewels, makeup, weapons. Those surfaces are always inscribed with the signs of social difference: as they are brought into visibility, bodies become classed, racialized, sexed and gendered in particular ways. Some bodies are constituted as human in this process and others as less-than-human, from which follows “the political, economic, cultural, and social exploitation of visible human difference” (Weheliye 2014, 6), as well as rich and diverse traditions of picturing bodies differently.

This chapter focusses on seeing bodies when digital data is harvested from bodies and is processed into images in urban contexts. As the technologies which visualize bodies are more and more often digital, so what it is to see bodies and to be seen as a body is shifting. The chapter points to the ways in which many digitally-mediated images make no reference to the traditions and practices just sketched all too briefly. Many digital images of bodies do not visualize corporeal epidermal surfaces on which signs of social difference are scripted. Instead, bodies are made visible in other ways. The chapter speculates on what some of these new bodies might look like, and not only what they might look like, but what it might be like to look at them. It does this by briefly exploring how bodies are pictured in urban spaces, as the cities in which so many human and other bodies live are also increasingly mediated digitally.

So, this chapter speculates on the reconfiguration of embodiment in digitally mediated cities. For Kember and Zylinska, mediation is “a multiagential force that incorporates humans and machines, technologies and users, in an ongoing process of becoming-with” (2012, 40). The chapter is interested in what new kinds of becoming-with might be emerging in cities saturated with digital technologies, particularly digital cameras of many kinds. If mediation is “a key trope for understanding and articulating our being in, and becoming with, the technological world, our emergence and ways of interacting with it, as well as the acts and processes of temporarily stabilizing the world into media, agents, relations, and networks” (Kember and Zylinska 2012, xv), then these beings, becomings, acts and processes must always reckon with “the insolvent place of the body in relation to new media technologies” (Munster 2006, 12). This chapter focuses specifically on how the digital devices and flows of data that now saturate so many cities are mediating the visualization of human embodiment.

While much attention has been given to the apparently nonhuman agency of software and code in digitally-mediated cities, nonetheless human bodies are intimately entangled with urban digital tech (Barns 2020). Bodies hold, carry and touch smartphones, tablets, laptops, gaming devices, eBook readers, smartwatches and fitness trackers; bodies trigger sensors, cameras, lidars and radars. Embodiments emerge in relation to these devices as much as they do in relation to other urban things like buildings, weather, crowds and signage. Hence we can say that bodies in many city spaces now simultaneously occupy both material urban spaces and digital, often onscreen, environments. This is what has been called the "'doubly digital' quality of contemporary media" (Moores 2014, 204): there is an intimate relation between the experience of being in online environments and the corporeality which enacts that experience. A body tapping or swiping on a smartphone is both in a street, say, and it is also in the virtual space of the screen. This doubling is not a merging or synthesis but what Munster calls a "a kind of graft, which is an unequivocal mark of connection and difference" between the fleshy body and virtual body (2006, 23): "New media entice bodies to venture towards incorporeal flows of information and combine, in convergent and divergent ways, the capacities and functions of carbon materialities with those of information flows" (Munster 2006, 23). This differential combining is what Munster (2006, 25) terms "digital embodiment".

Munster suggests that "our bodies are immanently open to these kinds of technically symbiotic transformations" (2006, 24–25). She describes how "in many information interfaces [...] the embodied self is forced into close proximity with itself as a dematerialized representation via the cursor, the feedback of virtual and actual gesture in immersive environments or bandwidth, and sensory compression in online interaction" (Munster, 2006, 25). Embodiment itself shifts as it is mediated by different technologies. Discussions of self-tracking are a pertinent example of this symbiosis (Lupton 2016; Neff and Nafus 2016). There are different kinds of digital self-tracking devices, but most bodies that run in city streets and parks will be wearing one, usually in a watch, wristband, or phone, which will be calculating data such as the route, distance, and speed of the running body, its heart rate and calorie consumption and the oxygen in its blood. A lot of people walking and cycling in cities use digital devices to track their activity too.

As Deborah Lupton (2016) says, this tracking can be understood as producing a particular kind of embodiment. Certain elements of embodied experiences are extracted as digital data and, importantly, are displayed on the device's interface. This data display then mediates further bodily activity. According to Kristensen and Ruckenstein, that data is monitored and evaluated by the self-tracker for whom "the use of metrics resonates with whom they aspire to become": body becomes test-site on which data-driven experiments are conducted, and data and corporeality thus "co-evolve" (2018, 3637). The work of Matthew Wilson (2011; 2014) offers a number of other examples of digital technologies co-constituting forms of embodiment in cities: bodies learning to respond to the auditory command for connection announced by the ringtones of the earliest mobile phones, or learning to notice the specific things in city streets that can be tagged by a handheld street survey device. The sensibilities of fleshy bodies are thus mediated by their engagements with digital devices. Hands learn new gestures, bodies learn new speeds and new forms of comportment, eyes establish new sensitivities to colour, and these perform new ways of moving, seeing and feeling. To quote Munster again, "the sensory and contingent plane of living bodies is doubled and variably reconfigured through computational schemas" (2006, 81).

Much of the data that mediates digital embodiment is materialized as images: the self-tracker interface, for example. Data that become images of various kinds are a significant part of the immense streams of digital data flowing through and beyond cities are images of cities and their inhabitants (McQuire 2016). Cities are full of people, looking at images of themselves and other people, on digital screens, in that city or somewhere else. Created and processed, distributed and redistributed, some of that data will materialize as photorealistic advertisements or news videos or selfies. Some will be converted into data visualizations: a dot on a map, a node in a network, a category in a graph. These many image types are seen variously in their turn, studied, glanced at, swiped through on a smartphone screen (Krajina 2014; Rose and Willis 2019) or a vehicle's GPS screen or reviewed and acted on in urban transport control centre (Kitchin, Maalsen, and McArdle 2016; Mattern 2015; Marvin and Luque-Ayala 2017). Made somewhere and uploaded, travelling somewhere else, being reconfigured, analysed, or multiplied, and then being downloaded, perhaps to the same device, perhaps to many others, in the same place or elsewhere: these are the distributed circuits of many forms of visual media now (Browne 2015; Casetti 2015; Munster and MacKenzie 2019). And at the interface between a digital image and its viewers, particular kinds of bodies coagulate. This chapter explores some of those coagulations.

Cities, data and seeing bodies

Most of the attention given in urban studies to digital mediation has focused on the "smart city" and its flows of big digital data (see for example Greenfield 2013; Hollands 2008; 2016; Karvonen, Cugurullo, and Caprotti 2018; Kitchin 2014; Marvin, Luque-Ayala, and McFarlane 2016). The term "smart" refers to the use of digital data to improve urban governance. For example, city authorities and commercial providers suggest that smart policies and technologies can enhance environmental sustainability by enabling the more efficient use of resources, especially energy and water; or that traffic flow or air pollution can be improved by using real-time environmental data; or that economic growth can be increased by innovating new products and markets based on digital data. Attention has also been given to the range of commercial platforms that collate and integrate urban data, such as AirBnB, Uber, and Mobike (Barns 2020). Social media platforms have been given less attention by urban scholarship but clearly also play a significant role in mediating urban experience (see for example Boy and Uitermark 2017; Leszczynski 2019a). As various feminist critics have pointed out, however, most of this work focuses on the agency of digital infrastructure and the political economy of its data extraction (Leszczynski 2019b; Rose 2017). Inspired by decades of intersectional feminist work, this chapter rather asks how to see bodies in this context of smart and platform urbanism.

This is an important question to ask since much of the data that circulates through urban spaces is about bodies. Corporations and city authorities produce, analyse and visualize, and act upon data about populations, citizens, commuters, taxpayers, users, residents, and consumers, among others. Residents and tourists picture and look at selfies, influencer videos, video chats, gifs; on Whatsapp, Instagram, and TikTok. Neighbourhood groups and activists use social media to share, witness and organize (Elwood 2020; Rodgers, this collection). And much of the data that circulates through and about cities materializes as images on some kind of screen, showing and being seen by specific bodies. This chapter approaches these images as a particular instance of how data mediates the constitution of embodiment.

This visual mediation can be described as ways of seeing which are co-constituted by the affordances of technologies, the social practices in which those technologies are embedded, their discursive framings and affective dispositions (see Gordon, 2010; Otter, 2008; Wilson, 2014). The phrase "ways of seeing" is John Berger's (1972). The chapter follows his use of the term to refer not only to what is seen but also to the body doing a particular kind of visualizing. Berger, like so much more recent critical visual scholarship, was concerned not only with how things are pictured but also with how specific kinds of picturing constitute particular kinds of viewers. He was also attentive to (some of) the ways in which the relation between seer and seen is always riddled with power. If visual images "body forth" corporeality (Copier and Steinbock 2018, 924), these are constituted as specific kinds of bodies, in particular relations. As this chapter's introduction noted, ways of seeing enact classed, racialised, sexed, and gendered relations; they position both seer and seen in those terms.

Many digital images mimic visual genres with long histories, which shape their making and their viewing. Maps generated by Geographical Information Systems may use the same data and be seen in the same way as earlier hand-drawn maps of census data, for example; three dimensional digital models of urban morphology may be used in the same way as architectural models. But these ways of seeing and constituting bodies are being joined by other ways of seeing other forms of embodiment which are caught up with the specific dynamics of data circulation. Digitally-mediated ways of seeing bodies are emerging that render bodies visible in different ways. Specifically, the chapter suggests that *animatic* embodiment is part of the digital refiguration of how urban bodies become visible.

To make this argument, it draws on Deborah Levitt's (2018) account of animation. Animation brings things to life. For Levitt (2018, 1), animation is "the dominant medium of our time", moving us from "questions about ontology, category, and being to ones of appearance, metamorphosis, and affect" (Levitt 2018, 2). Animation is both a medium and a contemporary cultural condition, according to Levitt. Its logic is not based on correspondence with a real; instead, animations envision metamorphosis, erasure, and resurrection rather than ontological presence. Schematically, the chapter argues that the organization of bodies in cities is also shifting between representational and animatic forms.

Animation, softimages and urban bodies

For Levitt, "new forms of life and vitality emerge at the spectator-screen intersection as this transforms over time" (2018, 3). While not directly determined by technological changes in the spectator-screen intersection, animatic life is enabled by some of the specific dynamics of digital images. It is important therefore to say a little more about how digital visual imagery is different from previous image forms. Hoelzl and Marie's discussion of what they term the "softimage" is useful here:

> As a program, the image, while still appearing as a geometrical projection on our screens, is inextricably mixed up with the data (physical and digital) and the continuous processing of data. What was supposed to be a solid representation of a solid world based on the sound principle of geometric projection (our operational mode for centuries), a *hard* image as it were, is revealed to be something totally different, ubiquitous, infinitely adaptable and adaptive and intrinsically merged with software: a *softimage*. (2015, 7)

A defining element of that merger is the software that allows digital images to be networked images (Munster and MacKenzie 2019; Rubinstein and Sluis 2008). The shift from analogue to digital photography, for example, was enabled not only by digital cameras but also by increasingly seamless connections between cameras, other viewing devices like computers and then phones and social media platforms. Digital cameras not only allow numbers of photographs to increase exponentially; those images and others become "ubiquitous" through their extensive circulation (Hand 2012). They are also constantly processed, in their making, their mobility, their multiplicity, and their materializations (see the introduction to this collection; Rose 2016). Munster and MacKenzie (2019) describe encounters with the resulting aggregate of innumerable images as "platform seeing".

The circulations of social media are saturated with images, of course, and design professionals create elaborate computer-generated images via a global division of labour (Chung 2018; Rose, Degen, and Melhuish 2014). Images appear on multiple screens, in different forms, at different sites. They are shared and favourited, liked or deleted, copied and posted, circulated and recirculated. There are nonhuman agencies at play here too. Much visual data is processed algorithmically; indeed, Steyerl suggests that "contemporary perception is machinic to a large degree. The spectrum of human vision covers only a tiny part of it" (2017, 47; and see Munster and MacKenzie 2019). The speed of image production, processing, and circulation is enacting a shift to "something more akin to live transmission" (Rubinstein and Sluis 2008, 22). The speed and scale of image production and analysis suggests not so much a network as a vast "stratified constellation of technical memory matter, composed of resources that shape political and cultural imaginaries [...] with depth, height, scale, extensiveness and duration [...] moving in different directions [...] Its forms may change and its content migrate, accruing or shedding textures in the process" (Withers 2015, 17).

The circulation of online images is encouraged by the centrality of sharing to social media generally (John 2016). In this way, data generated by digital devices "not only become [a body's] prosthetics but extend it into a network of *other bodies and objects*" (Lupton, 2016, 71, emphasis added; and see Gordon and Silva, 2011). Lupton points out that in discussions about self-tracking, for example, "it is contended that one can achieve the optimal self more quickly as part of a participatory culture" (2016, 133). Thus all self-tracking devices and apps have the facility to share the digital trace of a run or a bike ride on social media, often by posting a map of the route taken. Indeed, data about embodied activity, once digitized and uploaded, can become part of a quite different form of digital embodiment. Digital data about bodies circulates along complex and diverse paths (see also Casetti 2015; Munster 2013; Rubinstein and Sluis 2008). Data from social media posts and from apps that track physical activity and bodily health are available not only to the app users. They will also be shared with the app developer, who may well collate that data and sell it on, for medical research perhaps or to medical device manufacturers, so they can refine or sell their products (Lupton 2016, 119). Much attention has been paid to how platforms aggregate and recombine data, deploying machine learning to configure patterns that may align – or may not – with historically embedded regimes of social differentiation. What Cheney-Lippold (2017) calls "algorithmic identity" may have very little to do with class, race or gender, though the databases from which machines learn very often have those forms of differentiation embedded (Benjamin 2019; Noble 2018).

Softimages, then, are networked, processed, and distributed. So how are bodies visually mediated by softimages? The chapter will now sketch two visual regimes: the *representational* and *animatic*. Representational and animatic ways of seeing are not completely distinct, nor do they map neatly onto different technologies (as Levitt [2018] notes). Technologies, practices, discourses and affects all contribute to each visual regime. As the chapter has already noted, many digital images mimic the appearance of non-digital images and are looked at in ways that those have long been looked at, as representations of particular forms of urban life: closed circuit television (CCTV) footage taken with digital cameras, for example, can be seen in the same way as video taped CCTV. *Animatic* refigurations of urban environments and inhabitants, however, invite a different way of seeing – emergent, distributed, transformational – whether they are hand-drawn cartoons of urban superheroes or real-time maps of Twitter sentiment.

Seeing bodies representationally

The chapter now indicates how bodies continue to be visualized through *representational* ways of seeing. Barad, Kember and Zylinska (2012, 31) describe representationalism as the conviction that what is represented exists independent of all practices of representation. Representational ways of seeing assume that there is a real that images – no matter how selective and distorted – re-present to the viewer. In terms of visuality, cinema and photography have both been understood as media which are themselves representational but are also parts of a wider representational visual culture which Levitt calls the "cinematic regime" composed of "light, the machine, and an analytic eye" (2018, 12). In Levitt's account, as in that of many others, this is not the only visuality enacted by cinema. However, the dominant representational cinematic regime sees bodies in particular ways. What we see when we see bodies cinematically are "autonomous, massy entities" (Levitt 2018, 12) that refer to a pro-filmic real. Twentieth-century traditions of urban documentary photography, for example, are cinematic.

For Levitt, then, the representational cinematic regime entails the seeing of massy bodies by an analytic eye. In terms of thinking about bodies in urban space, seeing bodies as fleshy entities to be analysed is part of the "productive, biopolitical dimensions of cinema in the discourse of reflection, representation, and reality" (Levitt 2018, 11). That is, the representational cinematic regime is aligned with Foucauldian biopower (Levitt 2018, 28). This is a crucial step in understanding representational regimes of urban visuality. Representational urban visuality sees bodies as masses: as epidermal volumes with surfaces that can be analysed and categorized.

In biopolitical regimes, ways of seeing analyse bodies through discursive codes and in so doing, enact a range of social categories such as sex, gender and race (as well as distinguish between bodies that are human and not [Butler, 2007]). Feminist, critical race, and crip scholarship have been particularly attentive to this process. From hooks (1992) and Bordo (1993) to Weheliye (2014), Browne (2015) and Benjamin (2019), the ways in which bodies' surfaces are marked with, and interpreted through, visual signs of social difference have been described. And this analytical, biopolitical way of seeing continues in the digitally-mediated city. "Visual technologies and racial taxonomies fashion each other" (Benjamin 2019, 99). Corporeal bodies are rendered as the real surface on which particular social categories are visible. Facial "recognition" software can identify the gender, sex, race and even sexuality of bodies, we are told (though their failures are also regularly reported [Leszczynski, 2019b]). The notion of "recognition" here exemplifies the representational regime in which a reality is there to be recognized; it both assumes a real and obscures the process by which algorithms are trained to produce that "recognition" by human operators (Benjamin 2019). We might also consider the ways that big urban data is converted into conventional demographic or geospatial categories and then into dashboard displays (Ruppert 2012). From mapping population distributions to filming commuter crowds, visual techniques for seeing city inhabitants became part of understanding and governing urban bodies.

Understanding the digital mediation of bodies and cities as representational produces a particular form of critique. Much criticism of the algorithmic production of urban code/space (Kitchin and Dodge 2011) is based on the representationalist assertion of a "radical incommensurability between embodied and represented life" (Agostinho 2018, 143). Assuming that incommensurability between a real and its representation, this way of seeing is challenged because it "renders some things more visible than others, yielding new parameters of visibility that determine who or what dis/appears" (Agostinho 2018, 132). Thus planners' and developers' digital images of new urban developments are criticized for their inaccuracy: too clean, too sunny, too wealthy. Visualizations of smart cities glowing with seamless flows of data are criticized because actual data flows are "buggy, brittle and hackable" (Kitchin 2014, 1). In this regime, the who and the what exist before they are brought into, or excluded from, visibility.

It is this assumption of a pre-existing real that animatic ways of seeing abandon. Levitt's (2018) account of the representational cinematic regime and the animatic apparatus is careful not to assume a recent and complete transition from a film-based regime to a software-based apparatus. She is also careful not describe specific visual technologies as either representational or animatory. As she notes, films can be animated and animations can be representational. She does suggest, however, that a new technological and discursive dispositif is emerging in which "it doesn't matter so much what life is, as rather what you can do with it" (Levitt 2018, 20). Like Clough (2018), she references the life sciences where bodies are increasingly understood less as corporeal masses and more and more as information which – ever since gene-splicing, DNA mapping and transgendered bodies became possible – can be transformed and recombined as part of a wider contemporary cultural shift.

Animatic embodiment

The mediation of bodies into digital images happens in very many everyday urban contexts, as the introduction to this chapter noted. And while some of that mediation may take representational forms, animation also occurs, more and more often. Deleuze (1992) famously argued that the biopolitical was being displaced by what he termed "societies of control". In societies of control, the social is constituted through constantly mediated data and there is no sense of an incommensurability between that data and a real. Incommensurabilities of various kinds remain, but are better thought of as glitches within and between different data flows, as Leszczynski (2019b) argues.

If animation is a cultural condition, then one of its central dynamics is the circulation and recombination of data: "[B]ig data sorting that is designed to collate seemingly unrelated sets with the intention of producing novel relations" (Clough 2018, 107). Emergent patterns within and between data replace correspondence between image and reality and between sign and category. As Rouvroy points out, this algorithmic logic "spares [humans] the burden and responsibility of transcribing, interpreting and evaluating events of world. It spares them the meaning-making processes of transcription or representation" (2013, 143). Animatic embodiment in a smart city is configured by such emergent, algorithmic constellations of data. The distinction between the real corporeal body and its appearance in the network no longer holds. "Embodiment cannot be contained within the organic skin" when traces of digital data now so fundamentally compose what has to be called the posthuman (Clough 2018, xxxii). Bodies are turned into data and the algorithmic analysis of big data produces new corporeal entities (Rose et al. 2020): "No longer bearing a primarily representative function, we see the power of algorithmic vision today is its ability to craft an algorithmic subject that is always open to contingency" (Uliasz 2020, 7).

Such algorithmic subjects already inhabit the aggregate constellation of platform seeing. Computer game avatars have been joined by the deep fakes that circulate on social media; films routinely use motion capture technology; 3D models of heritage sites offer impossible viewpoints; augmented reality apps offer a disjunctural "visual cacophony bound together by a point cloud" (Uricchio 2011, 30). Digital visual effects saturate the visual field, and "paradoxically, as images of faces are amassed in digital form, the face recedes from our sight" (Uliasz 2020, 3). Much could be said about these various renderings but here my focus is on urban corporealities and the doubly digital articulation of digital embodiment.

Gabrys's (2014) discussion of smart citizenship is suggestive here. Gabrys explores the notion of smart citizenship a smart city design proposal projects and suggests that its production of digital data produces smart citizens as "ambividuals": citizen-subjects whose emergence is contingent on events, articulated through the technologies and practices of computational urbanism. Citizenship, she suggests, can thus become less an individualized subjectivity and more a digitally-mediated distribution, configuring "ambient and malleable urban operators that are expressions of computer environments" (Gabrys 2014, 42–43). So what do such "urban operators" look like? How might animated embodiment in a smart city appear? If the bodies in digitally-mediated cities are not all configured as the "body-as-organism", the body that is pictured in the cinematic regime, the massy body bounded by skin, if some forms of embodiment no longer look like that, how are they visible? How are they seen, and by what kinds of viewing?

Levitt's (2018) response is instructive. She not only looks at different things, but also looks at them differently. She views and reads a range of films and books not as representations but as animations. In animations, says Levitt, things are only erased or mutated or resurrected; things are not categorized but transformed. Her methodology is therefore to read animations not for their correspondences, references and presences, but rather for their transformative generation of novel sensations. Animation must be seen as an enactment; it exemplifies, not signifies; it must be looked at less for what and more for how. She turns her gaze onto dolls, automata, cyborgs, and robots and other urban inhabitants that have long figured in films as not-quite-human. These are figures, she says, that continue to look human enough for viewers to relate to them but are also different enough to suggest other forms of lively animation. They retain "just enough resemblance so that its potentials – if not the extent of its 'dimensions, depths, and distances' – become graspable" (Levitt 2018, 51). This potential is the animatory vitality of recombinant invention and other forms of emergent embodiment.

What then might become visible if we look for animatory bodies in contemporary urban spaces? Well, we might turn to autonomous technologies like driverless cars (Hind, this volume). Or perhaps to delivery robots, trundling alongside other pedestrians. Delivery robots have inhabited the streets of the UK city of Milton Keynes for some time now, for example. They are used by several supermarkets to deliver shopping and seem to have been welcomed as charming if puzzling additions to the city streets. A small sub-genre of YouTube videos has appeared, filming the robots and humans in Milton Keynes who pat them, talk to them, beckon them as if they were dogs, and watch them negotiate tricky kerbs. Several videos do nothing but observe the robots on the move. These videos are handheld, with no soundtrack, narrative arc, or interpretive commentary: they simply watch the delivery robots move. Humans seem unclear if robots are human-like or animal-like – but they definitely demand observation. Thus their automation is video-ed to mark something new and uncertain on the streets: something emergent. The digital video itself is bringing them into visibility, processing their presence just as the robots themselves constantly process sensor data in order to navigate. Without commentary, the videos render a new kind of urban embodied presence.

Robots are not the only kind of animated embodiments in digitally mediated cities however. Many other kinds of bodies are emerging from analyses of the data generated by urban sensors. These are the "data doubles" that shadow corporeal bodies in city streets (Cheney-Lippold 2017). Malleable and ambient, these are masses convened from data processed in near real-time.

These sorts of bodies are not so visible through the representational ways of seeing established for analogue photography and film. They become untethered from the many bounded bodies of a crowd, and are reconstituted as pulsing colours in animated data visualizations. Think for example of the mini-gif genre of the "breathing city", in which the population density of a city is visualized over the course of a day, pulsing in and out, it feels, as commuters arrive and leave each day. Bodies here are rendered as mobile geolocations, components of a city-scale animation. The geolocation data generated by commuters becomes a different kind of body en masse, a city inhaling and exhaling.

But perhaps the most animatic body in digitally mediated cities is the body seeing animations. Rather than the cinematic regime's positioning of the analytical spectator as separate from the screen, animations tend to merge and exchange image space and body space (Levitt 2018, 83). The screen is now an interface and the viewer is reconfigured as the screen reconfigures. No longer a single point of view framed by perspectival techniques, the spectator becomes a constantly mobile point of view, decentred, zooming and hovering through an environment that seems to have no frame as it is swept up and around by the softimage (Elsaesser 2013; Uricchio 2011). Elsaesser (2013, 240) describes this unanchored viewing, tracking seamlessly through spaces from the nano to the planetary, as "the default value of digital vision" (and points to its nondigital precedents in a range of efforts to create convincing three-dimensional films).

As well as this hypermobility, however, there is a sense in which the animatory viewer is constantly being reconfigured. Here, the plethora of digital visualizations that now bring urban spaces into visibility are particularly relevant. Cities now are insistently visualized through multiple interfaces, in different formats, genres, and media. There is no single frame, no nest of scales, no coherent territory. Different images merge and blend and the same image reappears in different contexts. Rather like Steinbock's (2019) trans account of cinema, this is a way of seeing cities through cuts and layers, disjunctures and recombinations, mutations, and excrescences. References to a real become references to multiple reals become a seamless dissolution of one real into another becomes more visual flow, exemplified in videos evoking the smart city which seamlessly mutate between multiple types of visualizations of city spaces (Rose 2018). Cities are thus also constantly transformed in the "mixed-space effect" of animation (Levitt 2018, 68) and so too are their viewers. As digital images emerge and mutate, their viewers are "not consolidated in identity but rather [...] consist of roving populations of action in the network" (Halpern 2015, 240).

Re-refiguring digitally-mediated cities and bodies

So if we think about the digitally-mediated city and ask – what do its bodies look like? – this chapter has suggested that there are several ways to answer the question. Drawing on Levitt's (2018) work, the chapter has sketched a regime of *representation* and what it makes *visible*, and a regime of *animation* and how it makes things *live*. The chapter has discussed how bodies in the digitally-mediated city can be thought of representationally in terms of features distributed on epidermal surfaces, which represent certain pre-existing social categories and which can be analysed. Animatic bodies, on the other hand, are seen and sensed as constantly emergent, mobile, fluid, and mutating. These two configurations are not only constitutive of how embodiments are made visible but also of different regimes of seeing. Representational ways of seeing assume a pre-existing real which images re-present to a stable viewer who decodes their meaning. Animatic ways of seeing, on the other hand, assume a viewer as emergent, mobile, and processual as the data they see.

The question of power in this visual culture remains, as always. Representational and animatic regimes of urban embodiment have different accounts of power. In representational regimes, power consists of the ability to analyse, identify and label bodies and spaces; to misrepresent; to exclude and render invisible. The power of animatic regimes also rests in the power to analyse and define, but also to extract, transfigure, morph and assimilate corporealities and cities. Each requires different actions, engagements, resistances and ethics. Urban scholars often seek resistance to exclusion. But animation demands a different critique. If the animatic apparatus "reorders the self as data [...] then it is important to continue opening out the question of the self and its constituent, relational others in order to see what kinds of relations are facilitated and prohibited in the process and what consequences various enactments of relationality will have, for 'us' and 'the world' at large" (Kember and Zylinska 2012, 112). The challenge then is to calibrate potentials for other transformations and potentials (Clough 2018), and other recombinations (Rose 2017), in urban data circulations.

References

Agostinho, Daniela. 2018. "Chroma Key Dreams: Algorithmic Visibility, Fleshy Images and Scenes of Recognition." *Philosophy of Photography* 9 (2): 131–55.

Barns, Sarah. 2020. *Platform Urbanism: Negotiating Platform Ecosystems in Connected Cities*. Basingstoke: Palgrave MacmIllan.

Benjamin, Ruha. 2019. *Race After Technology: Abolitionist Tools for the New Jim Code*. Cambridge: Polity Press.

Berger, John. 1972. *Ways of Seeing*. London: BBC with Penguin.

Bordo, Susan. 1993. *Unbearable Weight: Feminism, Western Culture, and The Body.* Berkeley: University of California Press.

Boy, John D, and Justus Uitermark. 2017. "Reassembling the City through Instagram." *Transactions of the Institute of British Geographers* 42 (4): 612–24.

Browne, Simone. 2015. *Dark Matters: On the Surveillance of Blackness*. London: Duke University Press.

Butler, Judith. 2007. "Torture and the Ethics of Photography." *Environment and Planning D: Society and Space* 25 (6): 951–66.

Casetti, Francesco. 2015. *The Lumière Galaxy: Seven Key Words for the Cinema to Come*. New York: Columbia University Press.

Cheney-Lippold, John. 2017. *We Are Data: Algorithms and the Making of Our Digital Selves*. New York: New York University Press.

Chung, Hye Jean. 2018. *Media Heterotopias: Digital Effeccts and Material Labor in Global Production*. London: Duke University Press.

Clough, Patricia Ticineto. 2018. *The User Unconscious: Affect, Media and Measure*. London: University of Minnesota Press.

Copier, Laura, and Eliza Steinbock. 2018. "On Not Really Being There: Trans* Presence/Absence in *Dallas Buyers Club*." *Feminist Media Studies* 18 (5): 923–41.

Deleuze, Gilles. 1992. "Postscript on the Societies of Control." *October* 59: 3–7.

Edwards, Elizabeth, ed. 1992. *Anthropology and Photography 1860-1920*. London.

Elsaesser, Thomas. 2013. "The 'Return' of 3-D: On Some of the Logics and Genealogies of the Image in the Twenty-First Century." *Critical Inquiry* 39 (2): 217–46.

Elwood, Sarah. 2020. "Digital Geographies, Feminist Relationality, Black and Queer Code Studies: Thriving Otherwise." *Progress in Human Geography*, January, 030913251989973.

Gabrys, Jennifer. 2014. "Programming Environments: Environmentality and Citizen Sensing in the Smart City." *Environment and Planning D: Society and Space* 32 (1): 30–48.

Gordon, Eric. 2010. *The Urban Spectator: American Concept Cities from Kodak to Google*. 1st ed. Hanover, NH: Dartmouth College Press.

Gordon, Eric, and Adriana de Souza e Silva. 2011. *Net Locality: Why Location Matters in a Networked World*. Oxford: Wiley-Blackwell.

Greenfield, Adam. 2013. *Against the Smart City*. London: Do Projects.

Halpern, Orit. 2015. *Beautiful Data: A History of Vision and Reason Since 1945*. Durham, NC: Duke University Press.

Hand, Martin. 2012. *Ubiquitous Photography*. Cambridge: Polity Press.

Hoelzl, Ingrid, and Rémi Marie. 2015. *Softimage: Towards a New Theory of the Digital Image*. Bristol: Intellect Books.

Hollands, R. G. 2016. "Beyond the Corporate Smart City?: Glimpses of Other Possibilities of Smartness." In *Smart Urbanism: Utopian Vision or False Dawn*, edited by Simon Marvin, Andrés Luque-Ayala, and Colin McFarlane, 169–85. London: Routledge.

Hollands, Robert G. 2008. "Will the Real Smart City Please Stand Up?" *City* 12 (3): 303–20.

hooks, bell. 1992. *Black Looks: Race and Representation*. Boston: South End Press.

John, Nicholas. 2016. *The Age of Sharing*. Cambridge: Polity.

Karvonen, Andrew, Federico Cugurullo, and Federico Caprotti, eds. 2018. *Inside Smart Cities: Place, Politics and Urban Innovation*. 1st ed. London: Routledge.

Kember, Sarah, and Joanna Zylinska. 2012. *Life After New Media: Mediation as a Vital Process*. Cambridge, MA: MIT Press.

Kitchin, Rob. 2014. "The Real-Time City? Big Data and Smart Urbanism." *GeoJournal* 79 (1): 1–14.

Kitchin, Rob, and Martin Dodge. 2011. *Code/Space: Software and Everyday Life*. London: MIT Press.

Kitchin, Rob, Sophia Maalsen, and Gavin McArdle. 2016. "The Praxis and Politics of Building Urban Dashboards." *Geoforum* 77 (December): 93–101.

Krajina, Zlatan. 2014. *Negotiating the Mediated City: Everyday Encounters with Public Screens*. London: Routledge.

Kristensen, Dorthe Brogård, and Minna Ruckenstein. 2018. "Co-Evolving with Self-Tracking Technologies." *New Media & Society* 20 (10): 3624–40.

Leszczynski, Agnieszka. 2019a. "Platform Affects of Geolocation." *Geoforum* 107: 207–15.

———. 2019b. "Glitchy Vignettes of Platform Urbanism." *Environment and Planning D: Society and Space*, October, 026377581987872.

Levitt, Deborah. 2018. *The Animatic Apparatus: Animation, Vitality, and the Futures of the Image*. Winchester: Zero Books.

Lupton, Deborah. 2016. *The Quantified Self: A Sociology of Self-Tracking*. Cambridge: Polity Press.

Luque-Ayala, Andrés, and Simon Marvin. 2016. "The Maintenance of Urban Circulation: An Operational Logic of Infrastructural Control." *Environment and Planning D: Society and Space* 34 (2): 191–208.

Marvin, Simon, and Andrés Luque-Ayala. 2017. "Urban Operating Systems: Diagramming the City." *International Journal of Urban and Regional Research* 41 (1): 84–103.

Marvin, Simon, Andres Luque-Ayala, and Colin McFarlane, eds. 2016. *Smart Urbanism: Utopian Vision or False Dawn?* Abingdon: Routledge.

Mattern, Shannon. 2015. "Mission Control: A History of the Urban Dashboard." *Places Journal*. 2015. https://placesjournal.org/article/mission-control-a-history-of-the-urban-dashboard/.

McFarlane, Colin. 2016. "The Geographies of Urban Density: Topology, Politics and the City." *Progress in Human Geography* 40 (5): 629–48.

McQuire, Scott. 2016. *Geomedia: Networked Cities and the Future of Public Space*. Cambridge: Polity Press.

Moores, Shaun. 2014. "Digital Orientations: 'Ways of the Hand' and Practical Knowing in Media Uses and Other Manual Activities." *Mobile Media & Communication* 2 (2): 196–208.

Munster, Anna. 2006. *Materializing New Media: Embodiment in Information Aesthetics*. 1st ed. Hanover, NH: Dartmouth College Press, University Press of New England.

———. 2013. *An Aesthesia of Networks: Conjunctive Experience in Art and Technology*. London: MIT Press.

Munster, Anna, and Adrian MacKenzie. 2019. "Platform Seeing: Image Ensembles and Their Invisualities." *Theory, Culture & Society* 36 (5): 3–22.

Neff, Gina, and Dawn Nafus. 2016. *Self-Tracking*. Cambridge MA: MIT Press.

Noble, Safiya Umoja. 2018. *Algorithms of Oppression: How Search Engines Reinforce Racism*. New York: New York University Press.

Otter, Chris. 2008. *The Victorian Eye: A Political History of Light and Vision in Britain, 1800-1910*. London: University of Chicago Press.

Rettberg, Jill Walker. 2014. *Seeing Ourselves Through Technology: How We Use Selfies, Blogs and Wearable Devices to See and Shape Ourselves*. Basingstoke: Palgrave Macmillan.

Rose, Gillian. 2016. "Rethinking the Geographies of Cultural 'Objects' through Digital Technologies: Interface, Network and Friction." *Progress in Human Geography* 40 (3): 334–51.

———. 2017. "Posthuman Agency in the Digitally Mediated City: Exteriorization, Individuation, Reinvention." *Annals of the American Association of Geographers* 107 (4): 779–793.

———. 2018. "Look InsideTM: Visualising the Smart City." In *An Introduction to Geomedia: Spaces and Mobilities in Mediatized Worlds*, edited by Karin Fast, André Jansson, Johan Lindell, Linda Ryan Bengtsson, and Mekonnen Tesfahuney, 97–113. London: Routledge.

Rose, Gillian, Monica Degen, and Clare Melhuish. 2014. "Networks, Interfaces, and Computer-Generated Images: Learning from Digital Visualisations of Urban Redevelopment Projects." *Environment and Planning D: Society and Space* 32 (3): 386–403.

Rose, Gillian, Parvati Raghuram, Sophie Watson, and Edward Wigley. 2020. "Platform Urbanism, Smartphone Applications and Valuing Data in a Smart

City." *Transactions of the Institute of British Geographers*. https://doi.org/10.1111/tran.12400/.

Rose, Gillian, and Alistair Willis. 2019. "Seeing the Smart City on Twitter: Colour and the Affective Territories of Becoming Smart." *Environment and Planning D: Society and Space* 37 (3): 411–27.

Rouvroy, Antoinette. 2013. "The End(s) of Critique: Data Behaviourism versus Due Process." In *Privacy, Due Process and the Computational Turn: The Philosophy of Law Meets the Philosophy of Technology*, edited by Mireille Hildebrandt and Katja de Vries, 143–168, Abingdon: Routledge.

Rubinstein, Daniel, and Katrina Sluis. 2008. "A Life More Photographic: Mapping the Networked Image." *Photographies* 1 (1): 9–28.

Ruppert, Evelyn. 2012. "Category." In *Inventive Methods: The Happening of the Social*, edited by Celia Lury and Nina Wakeford, 36–47. Abingdon: Routledge.

Steinbock, Eliza. 2019. *Shimmering Images: Trans Cinema, Embodiment, and the Aesthetics of Change*. London: Duke University Press.

Steyerl, Hito. 2017. *Duty Free Art: Art in the Age of Planetary Civil War*. London: Verso.

Uliasz, Rebecca. 2020. "Seeing like an Algorithm: Operative Images and Emergent Subjects." *AI & SOCIETY* Online First (September): 1–9.

Uricchio, William. 2011. "The Algorithmic Turn: Photosynth, Augmented Reality and the Changing Implications of the Image." *Visual Studies* 26 (1): 25–35.

Weheliye, Alexander G. 2014. *Habeas Viscus: Racializing Assemblages, Biopolitics, and Black Feminist Theories of the Human*. Durham NC: Duke University Press.

Wilson, Matthew W. 2011. "'Training the Eye': Formation of the Geocoding Subject." *Social and Cultural Geography* 12 (4): 357–76.

———. 2014. "Continuous Connectivity, Handheld Computers, and Mobile Spatial Knowledge." *Environment and Planning D: Society and Space* 32 (3): 535–55.

Withers, Deborah. 2015. *Feminism, Digital Culture and the Politics of Transmission*. London: Rowman & Littlefield International.

About the author

Gillian Rose is Professor of Human Geography at the University of Oxford and a Fellow of the British Academy and of the Academy of Social Sciences. She is the author of *Feminism and Geography* (Polity, 1993), *Doing Family Photography* (Ashgate, 2010), *The New Urban Aesthetic: Digital Experiences of Urban Change* written with Monica Degen (Bloomsbury, 2022) and *Visual Methodologies* (Sage, fifth edition 2022), as well as many papers on images, visualising technologies and ways of seeing in urban, domestic and archival spaces. Her current research interests focus on contemporary digital visual culture.

7. Speculative Digital Visualization as Research Strategy: City Building through Mobile and Wearable Camera Footage

Asli Duru

Abstract
Thinking through the affordances of everyday video technologies, this chapter reflects on a speculative research methodology and its processes of shooting, editing, and sharing of videographic knowledge on ordinary violence in Istanbul. Memories and perceptions of what violence looks like, how it feels, and how it affects the micro-constitution of everyday life is visualized through participant-generated smartphone and wearable camera footage, which resulted in a research video as a speculative sense-making platform where visual friction, silence and confusion in raw footage are expressed in filmic colour, sound and rhythm. The chapter makes the case for speculation as a strategy to sense what world-building opportunities might appear once the will to certainty is consciously done away with.

Keywords: ordinary violence, speculation, filmic ethnography, Istanbul

Introduction

This chapter presents a methodological ethnography, and specifically focusses on speculation as a critical feminist methodological strategy that guided the making of the research video *A Walk Down the Shore*. The video is based on the socio-materiality of smartphone images, wearable camera footage and film-editing as specific forms of urban digital data that afford unique ways of articulating the spatial experience and political meanings of

Rose, G. (ed.), *Seeing the City Digitally. Processing Urban Space and Time.*
Taylor & Francis Group, 2022
DOI 10.5117/9789463727037_CH07

everyday urban violence in Istanbul. Set in the Maltepe-Kadikoy coastal strip, the study addresses the social, material and symbolic modalities of violence felt, experienced and remembered during the course of mass (third to fifth generation) redevelopment ongoing in the nearby neighbourhoods roughly since the 1950s but intensely in the last decade. Using participant-generated mobile photography, video and visual elicitation methods, it interrogates the role of memories and embodiment in enabling and disabling individuals' conception of violence and wellbeing in the city; and the potential of everyday digital visual practices as methods in affective urban research.

In order to sense and explore the mediated, embodied geographies of human perception and embodiment of harmful encounters (and remedies), the study relies on memories and remembrance as the gateway into the narrative and movement-based im/mobilizations shaped by gender, age, class as well as by sound, light, and the presence/absence of human and non-human others including waste, smartphone coverage, and/or the camera. Conceptually, the research is also organized around the ordinary violence framework which involves the structural, slow, meta-, continued, habitual forms of aggression in urban spaces.

Ordinary violence prioritizes the micro-constitution of difference and actions and decisions taken/not-taken due to feelings of fear while moving and interacting with others in social space. The major reason for reframing existing approaches to violence around the ordinary is to activate ways of understanding violence as a mode of doing, perhaps in the simplest and most ordinary sense rather than as some thing, an event, or a series of "that which happens". Bringing the attention from the event to how violent encounters feel and look like, the argument centralizes the body as geographical scale and calls for a methodology that can both displace the common-sense references of violence as "violent event" and articulate the diversity of violent encounters as they are felt, remembered and visualized (Duru, 2019).

Many urban spaces are increasingly governed by new forms of corporate storytelling that more and more impose narratives around the smart, eco-friendly, and secure city. These narratives in practice refer to services and technologies that afford local governments more efficient and interconnected tools to address the blind spots in their surveillance, and in their financial and infrastructural networks. With narrative authority extending from humans to algorithms (not excluding the processes in which they are produced and governed), visual data plays a significant role in how urban stories are cast and mediated, and how new and old socio-spatial hierarchies and subjectivities are represented. In the case of Istanbul, corporo-political storytelling around growth, surveillance, and neo-imperialism are the themes of a prominent visual narrative that can be seen through the computer-generated imagery of mega projects like the disputed Istanbul Canal, the city's "Urban Information and Security System" (MOBESE), and the apparent export value of long aerial shots of the city in Turkey's booming soap opera industry. In this context of increased emphasis on maximizing Istanbul's sentience and visual intelligence in corporate, authoritarian, and automated spatial narratives, my intent is to discuss the potential of speculation as a visualizing strategy that can lead to a specific filmic urban imagery and intelligence that is "embodied, embedded, enacted, and extended" (Rowlands 2010, ix) among residents, digital technologies, and urban space.

A speculatively visual engagement with ordinary violence and subjective wellbeing in urban everyday environment is in dialogue with multiple theoretical and practical intersections between feminist storytelling (Pelc, Hasan, and Mollen 2020; Toupin 2018) and creative (Harper 2014; Skains 2018), practice-based (Chamaa 2017; Candy and Edmonds 2018), mixed-method (Hesse-Biber and Griffin 2015), and post-qualitative (Johansson 2016; Lather and Pierre 2013) methodologies. I suggest that, in addition to these, a discerning feature of speculative visualization is its non-antagonistic yet provocative, open and accepting stance to the possibility of knowledge that does not rely on the visual environment as a readily available domain to "fix or arrest meaning" (Oxman 2010, 77). Drawing on Barthes, Oxman elaborates on the scientific imposition of the visual as sign; that is, essentially as an object of knowledge with "identity and closure" compelled to bear an answer to what it refers to so as to "prevent it from trembling or becoming double or wandering" (Barthes 1973, 33, quoted in Oxman 2010, 77). Since the early days of photography the increased representational power of imaging sustains the scientific convention to detach visual outputs from questions of what visualization in and by itself is and does. A crucial dimension to this neglect also has to do with associating visualizations with the ethos, processes, and methodologies that succeeded in enabling and expanding the technologies that made high and higher resolution, malleable, editable, composite renderings possible. Yet, the infinite circulation of visual data on social media today is one way to observe that what visuals mean and do in social space is more expansive and full of unknowns that do not match the internal logics, technological workings, and the scientific narratives of the tools that enabled their presence. In line with this observation, my speculative methodology, rather, aims to mobilize visual practices and objects in order to get a sense of the existing and emergent worlds, hierarchies, and sensitivities that come alive through the interactions between visual practices, things and subjects.

I argue that speculative visualization as a methodology is as much about developing practices and tools as it is about repurposing technologies and sensitizing visual narratives in order to gain insight into the geographical meanings and relations generated by their circulation. Therefore, in its latest iteration, the video is a frictional narrative based on the interactive experience of knowing, navigating and accepting the liability of meaning to mute, wander, and become "grainy" through layers of audio, visual, textual data on how violence appears around us; what visibilities were at stake and for whom.

The pervasive use of urban imagery to represent "what already is" even though it is invisible or unimaginable to the eye is the departure point of this project. It will re-route itself towards a narrative on using visual methods in order to speculatively cast new, non-corporate, embodied and experiential stories of the city. In the following, I will elaborate on the process of making *A Walk by the Shore* so as to make a case for speculative visualization and to explain how it translated into staging the research, from the production of images to the editing of the film. In what follows, I will first ground and locate my approach to speculation as a critical visual strategy. I will then build the case around the making of the research video. I will pay attention to following the conventional research timeline from preparation through fieldwork and data processing as much as I can, but the narrative will rather present a processual flow organized in two sub-sections: theory building and creating surfaces. The conclusion will summarize the research and highlight the possible further routes and questions to seek meaningful and responsible, imaginative and situated concepts and frameworks for articulating visualization practices as a powerful means to interrogate spatial relations in cities and elsewhere.

Situating the speculative as a visualizing strategy

A growing body of feminist, technological, design and artistic interrogations have activated the speculative as a politically transformative logic of thinking and doing (Bivens 2018; Hackers of Resistance 2018; Leorke and Wood 2019). Feminist storytelling, speculative fiction as well as material, post-human, post-qualitative turns in academic and artistic research connect the web of ideas and practices that offer routes to take visualization practices seriously (Rose 2014). What these approaches share in common is an emphasis on critically engaging the possible and interconnected processes in which complex experiences of the human and non-human world, its ideas and emotions mediate.

At the core of post-qualitative interventions lies an anticipatory research process that invents and theorizes practices that interrogates the idea of a stable subject. Post-qualitative research processes are multi-directional; they grow rhizomatically, and can be followed and supported without the need for a schematization in order to thoroughly engage its dendrites. Knowledge is anticipated through the simultaneous interweaving of form and content (Duru 2020) in the absence of successive goals and prefigured aims. This however does not mean that process and practice-based methodologies resist meaning. Rather, the process involves a speculative dimension where the purpose is to multiply narrative opportunities and craft meaning through the openings, possibilities, forms, and interactions sensed and observed throughout the research process.

Research-creation (Basu and Macdonald 2007) is a cognate concept that emphasizes the material and embodied ties between research questions, context, and activities in the rhizomatic development of a methodology that is sensitized (Malinowska and Miller 2017) to the existing and possible presence of "entangled agencies" (Barad 2007, 22; 2014) and dispersed causal relations (Bennett 2010). The process involves a future-based aspect while maintaining a constant reference to situatedness; the here and now (Harding 1991) of the non/human elements and standpoints that shape the creative process and research practice. The speculative in this sense is constructed as an accountable thought process by situating existing and emergent variables through the constant spacing of time (here) and timing of space (now).

Speculative design requires special attention in order to situate the speculative specifically as a critical material and affective visualization approach. In connection with the emphasis on the speculative as situated, Dunne and Raby (n.d.) discuss the difference between techno-determinist futurism and speculative world building as a critical theoretical orientation and creative process. In articulating this difference, they emphasize the crafting of a narrative that is set in a possible world rather than the unfamiliar and shocking aspects of science fiction futures. The speculative in the first case is an "authorly" process that entrains both the familiar and unfamiliar to create an alternative reality where the meaning of technologies multiply, shift, and give rise to the discussion of their present publics and mediation (Auger 2013).

Design constitutes the foundational environment where the visual is co-constructed by the speculative. As technological objects, systems, roles, uses, and effects are conceived, the design and implementation of particular affordances for these objects are at the same time a process where the needs, use, and agency of possible users are imagined (Akrich 1992). In this sense, media design and evolving forms of mediation involve inherently anticipatory processes that require an actively speculative approach that can open up the political space where new and possible user subjectivities and experiences can be prefigured. Evolving political and corporate hack cultures, including tech-based abuse practices, provide examples of how this space exists and can be mis/used.

Speculative fiction has also become an increasingly popular genre that encourages fictional and radical imaginative approaches that invent new ways of living and dying and being and doing. Imarisha defines the genre as "whenever we try to envision a world without war, without violence, without prisons, without capitalism, we are engaging in speculative fiction" (2015, 3). Building on this definition, I argue that its privileging of radical imagination concentrates on two interrelated dimensions that provide the time-space for the speculative as an ethical and generative response to the inherently positivistic regimes of vision in research and other contexts. First, in both academic and artistic practice, speculative interventions have drawn attention to the repressed material ecology in which technologies appear, along with the bodies and structures that they interact with. Exploring this dimension leads the way to new questions around the material and embodied experience of technological worlds — how we perceive them, how they feel, and how these two questions are intertwined. Second, theoretical and artistic approaches to the embodied and material have also shifted the political vision from the past by integrating anticipatory emotions and ideas as a vital resource for knowing the present and its potentialities.

Both these material and anticipatory aspects emphasize the need to engage the messiness of the unknown unknowns (Allen 2011) and to expand empirical anxiety, towards developing questions, imaginaries, and scenarios that deal not only with "what is, but also what could be" (Leorke and Wood 2019, 63). The speculative is one strategy to mobilize the space conjured by the material, corporeal, and anticipatory dimensions of unstable, human/other agencies, and their relations. As a world-making tool, it situates the here and now of ambiguity (Bendon 2005), reframes questions accordingly and enables a route to articulate the subtle, future-sensitive, and embodied realities of diverse experiences of being and doing in a complex world.

In the specific context of visually exploring memories and perceptions of violence and wellbeing in Istanbul, incorporating speculation into research offers two key advantages. First, it activates the resources, maps and, where necessary, fills in the interstitial space of ambiguity that has been de-privileged in deterministic visualization practices. This aspect resonates especially with research-creation work already undertaken in creative and artistic fields that put academic, artistic, and empirical knowledge practices in dialogue with each other. Second, it allows for a wider, future-sensitive, and situated understanding of everyday visual practices and their meaning in existing and possible regimes of vision. Resisting the meaning of visual methods as a means of visual iteration of narrative phenomena, speculative visualization is "a delicate act of creation, something that requires time and effort to make realities and hold them steady for a moment against a background of flux and indeterminacy" (Law 2004, 10).

Haraway (2016) imagines the speculative as a methodology that calls for the invention of new perspectives and tools on the nature and extent of our connections with human and non-human others. In formulating this approach, she suggests the working of string figures which involves a simple loop that is transformed through the movements of fingers in a pattern that visually depicts an object or animal and is passed down through generations. String figures, as Haraway explains, work the same logic as speculative fabulation, speculative feminism, science fiction, and science fact in fabricating a story based on certain figures and patterns. In this sense, the story told and its media, in Haraway's phrasing, share a "response-ability" in their co-creation.

What can this metaphor offer in terms of research visualization practices? What can we make with evolving visual technologies in order to fabricate new and alternative stories of our human and non-human interactions? The metaphor of string figures illustrates the narrative process in which movement patterns creates a figure, this figure makes the story possible, and a different story can be told by undoing the figure, going back to the initial loop, and creating a new figure to tell a different story. The narrative precedes the figure, and embedded in this process is a political space that can be explored by the tools that appear through a speculative engagement with their affordances. Remembering that "it matters what stories we tell to tell other stories with" (Haraway 2016, 12), the stories, claims, and assumptions around our present and evolving technological tools and systems co-create and share the responsibility for the knowledge practices and worlds that we narrate by using them. In the following section, I will ground this argument by reviewing the process that helped me form a reality of using digital film and everyday smartphone mobile photography as visual methods. These methods went beyond the forensic use of what everyday visual practices – particularly smartphone images and footage – do and can mean as world building tools.

The making of the video

Geographies of ordinary violence can be mediated through the spatial affordances of editing software – namely through surfaces and layers of colour and sound as the styling tools – to express continuities, cracks, and sediments in the participant narratives. Moments of disruption and physical obstruction, indecision, and confusion spatialized on a filmic surface aim to distance viewers from a disciplined sense of flow and pace in the face of misplaced, ambiguous bodies, and subjectivities. The horizontal "timeline" of the software enabled layered projections of the bodies and objects that were marked, divided, and charged differently by memories. Long frames and slowness are also ways of negotiating flow (and obstruction) that are visual means of taking into account the discomfort of not knowing. The video, in this sense, is a sense-making platform where silence and confusion in the narratives lend their audible and rhythmic origin to spacing and visual friction as they are expanded in sediments of colour, transparency, and blur effects. The resulting landscape is marked by speculation and a loss of certainties where visual friction, silence, and confusion in raw footage expand in filmic layers of colour, double exposure, and blur effects recognizing the gap for (but not showing) possibilities of risk, hope, violence, and wellbeing.

Visual research methods, in the broadest sense, refer to knowledge practices that create visual objects to explore ways of seeing and been seen. As technologies diversify and become more widely available, and the range of experiences and affective affordances these new tools enable multiplies, their production of representational images still maintains close ties with the logic of indexicality or the "capturing" power of visual technological tools. Moving away from the assumption that a relationship exists between the object and its image, this video project took the alternative stance and sought meaningful and affective visual compositions of urban violence through the non-eventful, non-spectacular ordinary violence approach and aimed to articulate the minor, perceived, and embodied experience and memories of violent encounters with place, humans, and others in the urban social environment. In order to achieve this, the research process maintained a continued focus on re-theorizing what violence means and where it takes place in the absence of an eventful event in Istanbul's everyday settings and routines. This, in turn required a sensitized anticipation for the kind of violent encounters that almost feel like common sense, that is, for experiences and perceptions that are deemed normal compared to "violent events" where world-building through a forensic identification of victims, perpetrators, and aggressors is the convention.

Within such an approach, mobile photography walk-alongs and visual elicitation as research methods themselves became active means of conceptualizing ordinary violence in Istanbul. The walk-alongs, which took place in the Anatolian part of Istanbul with individual residents, were loosely organized. The research framework and questions were also made open to interpretation and discussed with each participant before, during, and after the walk-alongs. The methods and the tone of their articulation deepened the context emergent with and through the materialities of technical and social relations around ordinary violence and its visualization. This generative understanding of methods and context makes speculation a useful framework to advance questions of harm, safety, and place offering multiple possibilities to understand these from emergent perspectives.

Using speculative remembering and the felt knowledge of embodied emotions, the future-sensitive inquiry of the present city extended the political vision of violence to include the seemingly neutral, harmless object and place relations. In what follows, I will demonstrate this approach by elaborating two key streams that best represent the work and thought patterns of making *A Walk Down the Shore*. These processes enable a representative and transferable narrative on an extremely iterative work flow that aimed both to bring in materiality and leave space for ambiguity about how violence appears in everyday lives using memories, visualization and corporeality.

Building theory through practice

A Walk Down the Shore is a nineteen minute narrative consisting of eight episodes— theme units, pieces of video — that interweave participants' visualization of memories of violence and wellbeing through their present encounters with humans and artefacts in everyday life. It is driven by the assumption that visual practices are embedded in biographies. Life stories, memories, and perceptions of place are mediated through the video acts (Duru 2018). The digitally extended and embodied life of memories animates objects such as smartphones, selfie sticks, social profiles and action cameras. Hence the shifts in everyday visual technologies and practices imply the changing socio-materiality — lifespan, significance, and emplacement — of memories. Building on Haraway's (2016) work on speculative fabulation and everyday practices of storytelling, the video constructs a digital, corporeal, and speculative conceptual space to situate memories so as to (re)imagine and reconsider the different and hierarchical ways violence and wellbeing appear in everyday environments.

On a technical level, research creation in the study involved phases of walking, map visualization, and filmic iteration, not only as methods to generate data but as extended platforms of speculative exploration and theory in themselves (Chapman and Sawchuk 2012). The video is empirically based on the audio-visual material produced by fifteen residents living in redevelopment neighbourhoods in the Anatolian part of Istanbul. In the video, the past co-emerges with technologies of visibility and creates digital, site-specific, movement-based narratives of ordinary violence by sensitizing the participants and viewers towards habitual avoidances and preferences for certain ways of moving in the city due to fear and anxiety. The following excerpt from the video introduction describes further details on the process:

> Field research included wearable camera walks, smartphone photography, and visual elicitation meetings. Together with fifteen participants, our practice involved phases of ambulatory reflection, situated "mis-guidance" (Smith) and speculative visualization in order to explore the intimate geopolitics (Pain) of ordinary violence in Istanbul. We were loosely organized in the mobile photography tours and later in the visual elicitation meetings. Participants were informed that the project was about visualizing place-use and memories of violence. But they were free to interpret these relations within personal frameworks such as 'aggression', 'fear', 'safety', 'feeling good', and 'terror' in the city. During the walks, they took pictures of what they thought was somehow related to these notions in the immediate environments along the seafront in Anatolian Istanbul. In the visual elicitation meetings we reflected on the material and negotiated their various meanings and significance. During the elicitation meetings, I also asked participants whether they would share a particular image on social media, and if so, how they would "tag" or "caption" it. Hence, a textual layer of input along with oral and audio-visual was added to the material. Consent on the use and distribution of visuals was gradual and obtained on the basis of each audio-visual clip and photograph. (Duru 2020, 158-159)

I define the video as place-based "filmic ethnography" as opposed to "ethnographic film" since it is situated in speculative derivation. The filmic ethnography here is a thematic and aesthetic iteration of research material originally not intended for film. Non-textual, visual processing of visual data was one of the crucial aspects of the methodology, therefore it was only after data gathering and trying to find alternative visual forms, that video-editing emerged as the feasible and sound data interpretation platform in order to interrelate and create possible and relevant surfaces using the material. The major advantage of digital video over conventional text-based visual research outputs was: first, the availability of creating composite, layered narratives that enabled a place for not-knowing whenever it felt necessary; and second, the ability of the filmic to radically decentralize and yet complement text while also enabling a shift away from the authority and our reliance on the event. In other words, video editing became theory and theory became video that formed the reality of this filmic ethnography (Barad 2015). The speculative enabled the process to remain open to gaps, edits, and resampling (Lessig 2009).

Accordingly, filmic ethnography also underlines the difference between visual speculation, and speculative visualization for which representation is not intentional yet anticipated and acknowledged. This does not imply that speculation achieves making the more-than-representational visible. Rather, it means that the emphasis shifts from the representative power of research images towards theory building through dialogue (Peake 2016) with subjects and objects of knowledge, recognizing and making visible that alternative dialogues are probable and possible in the same process and with the same material.

Presenting everyday visual practices as methods meant a speculative methodology but not necessarily an artistic one, at least as I conceptualized this research. Still, in the elicitation meetings it became clear that most participants were involved in the project because the minimal design of research tasks had an artistic aspect. They described walking, chatting, and taking pictures as "optimistic" and "therapeutic" despite the discomforting content of our photography walks. This joyfully pessimistic and feel-good effect of mobile photography may be one way participants connected to the research process although artistic, aesthetic, or creative were not my identifiers of choice for the project. Yet interpreting mobile photography as an artistic practice speaks to a deeper concern of the project to resist the representational impulse, undo the assumption that variables must exist and to realize itself beyond utility, by visualizing and metaphor rather than showing and visibilizing. In other words, the anticipatory dimension to what we did, the risk of failure by deliberately "taking the wrong path" (Haraway 2016), and giving place to ambiguity and unknowing in visualizing memories of violence led to an excess space — an unstable epistemic moment (Lyotard, 1984) that flattened the fieldwork encounter into an existential exchange which felt creative and artistic. This excess space which is typically granted to creative practices is the place of the speculative and can offer a powerful means of decolonizing visualization practices from the hegemony of social scientific masculinity and epistemic aggression of arresting meaning in utilitarian regimes of vision. What is at stake here is not necessarily an antagonistic intention to fail these regimes but to establish responsible relations of visualization.

By the end of the fieldwork, I had no raw ethnographic material that I could process further in order to reduce and synthesize through clustering, sub-grouping and elimination. Whatever meaning existed was already curated, although it was fragmented and full of voids and personal myths. Data gathering, in that sense, resulted in a participant-curated visual collection (rather than raw data) that alerted me to style as an emergent sensory field that enhanced my interaction with "amateur" technologies and practices of everyday snaps which in turn led to further experimentation with video and remix.

Creating surfaces

Conceptual mapping was the first work surface that I created in order to map the tags and captions participants gave to their photos. This step modelled the initial coding step of qualitative analysis but allowed me to remain with participant generated material (tags and captions) as the first layer of codes. I then clustered these codes based on visual and narrative content including technical and material categories such as distance to object, position of the device, zoom effects and so on. I should note that creating clusters in this step felt and worked differently than clustering data in software-based analysis. This was because my practice of map-based clustering resulted in an alternative visualization of data in the form of schemes through a connecting strategy, as opposed to vertical processing by elimination in qualitative analysis software. This map-based horizontal connecting practice (rather than sub-grouping) draws on psychological "scheme theory" in which a scheme is defined as a psycho-social cluster of related pieces of information, knowledge, or memory that form a surface or a frame of reference for sense-making (Axelrod 2014). These visual clusters, or sense-making frames, formed the domain of speculatively exploring the derivations of the material. I worked a similar logic with wearable video footage where I created map-based conceptual schemes by time-stamping and connecting three pre-identified moments: entering the photograph setting, taking the picture, and exiting the photographic site for each participant video.

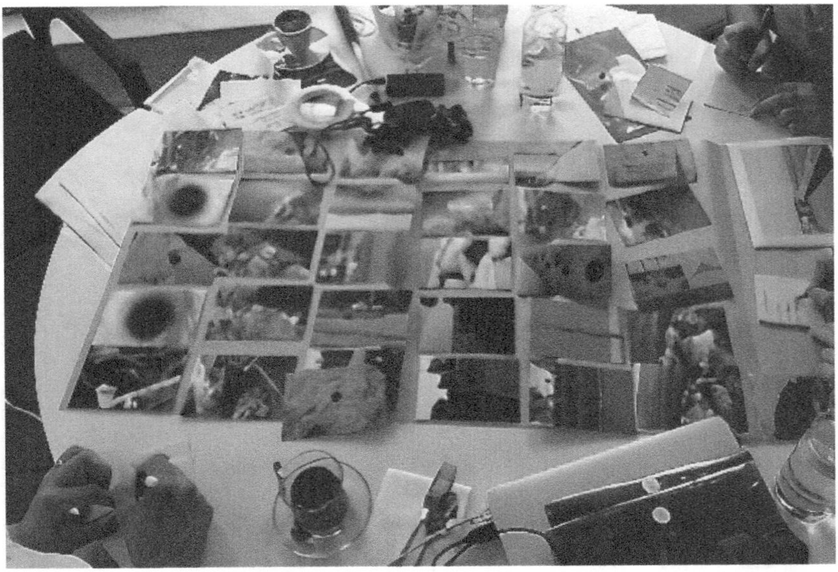

Figure 7.1. The elicitation meeting. Photo by author.

Figure 7.2. Post-fieldwork iteration process, working with a physical conceptual/timeline. Photo by author.

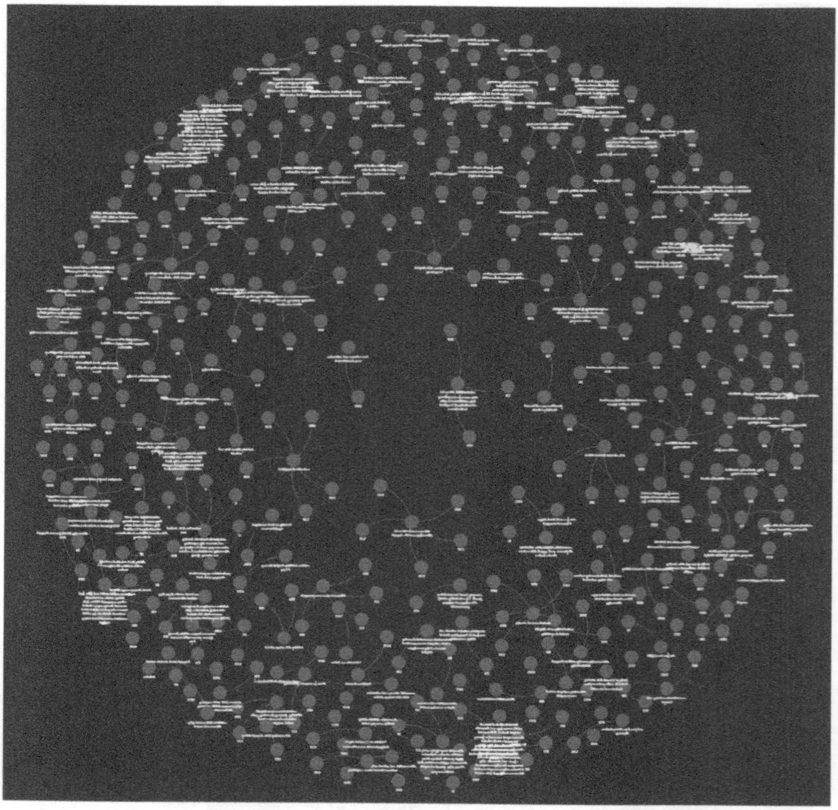

Figure 7.3. An alternative visual iteration in the form of a conceptual map that complemented the horizontal layering process. Made by author.

Figure 7.4. Screenshot from video cover. Video by author.

Based on the secondary visualization of data in conceptual map form, it was evident that gender made an important difference to participants' practices of visualizing harmful memories and experiences. Briefly, this was visually evident from the way female participants' visuals of an unknown object (e.g. garbage, broken beer bottles, unattended package) showed longer distance-to-object than did male participants'. Women's conception of what constitutes a landscape or portrait photo also reflected a sense of depth and focus on multiple elements in a setting, with longer captions than male participants' focus on objects and words rather than sentences when captioning. The second key difference that affected mobile visual practices was age. Both age and gender made a difference in participants' self-consciousness and fear of wearing and/or working with camera in public (Duru 2018). Vertical/horizontal positioning of the phone, feeling motivated for social posting, embodiment of visual tech and textual interpretation of images were, hence, meaningful in concluding that everyday visual technology and visual objects like snaps mean and do different things for particular subjects.

Remembering that the map is a way-finding tool but not the territory, the next phase involved alerting my senses to the stylization of data as a means of building a world of emotional nuances, gaps, sensations, and visibilities in the audio-visual narratives. Video-editing allowed me to continue my horizontal practice of connecting information and topographical strategy of creating surface, layering the various fragments and objects of knowledge into a consistent narrative. Consistency here draws on feminist objectivity and taking "data driven-ness" seriously, even when data itself is too intangible or non-existent for anyone beyond the participant and sometimes for the participant herself too. I argue that this is where letting style breathe through the affordances of the editing platform or the smartphone for expression (both participants and researcher) rather than editing, writing, and note-taking for representation co-emerges together with the choice of research tools as a crucial dimension of critical knowledge production.

The epistemological departure from "exposure" towards a data-driven navigation of style and "expression" had its own setbacks, such as the fine line between denying knowledge and making visible the denial of the assumption of knowing. In *Passing Drama*, video-artist Angela Melitopoulos (1999) engages similar concerns over the materiality of video and elaborates on the medium's philosophical relation to time, memory, geography, and subjectivity. The video-essay deals with the memory of political refugees deported from Turkey to Greece and later to Germany, many of whom experienced forced labour under the Nazi regime and repeated displacements thereafter. The film renders remembrance and forgetting as rhythmic structures through the interweaving of collective memory, subjectivity, and time. The moving image is linked to movement in space — a method of research and image-production that maps geography and psychology onto each other to reveal a different political imaginary (Lazzarato 2019), which helped me engage in a two-way dialogue between theory and practice while cutting, stitching, and layering the smartphone snaps and videos.

In a similar vein, in the experimental science fiction film-essay *Mnemophrenia*, Eirini Konstantinidou (2019) uses virtual reality (VR) as a metaphor for memory and envisions a future society where film achieves a complete simulation of reality, and artificial memories are generated that take on a life of their own. The two-way dialogue between theory and practice in these examples interrogate the visual technologies that produce the narratives enabling a speculative space where the physicality of the tools embody the possibilities in this space.

Similar object and body relations and socio-material connections across data and style were revealed in my filmic workspace. The past extended into the present, inhabiting objects and bodies, travelling through technologies. The narrative is fragmented and ambiguous. Ambiguity has volume; it is creative and it does things, enables and disables others to breath and to relate to others' atmospheres. Pressurizing it displaces those (ideas, agencies, experiences) who lack space and expression.

Many iterations surfaced through the horizontal workspace of the editing software which proved the "generative acts" of methods (Davies 2004, 26). Hence, transferring memory visualizations into video was by no means a straightforward process. Thinking through the materiality, affordances and ethos of these tools was a complex exercise in terms of reviewing the research material and questions, while at the same time re-assessing the dissemination and impact implications of the end product. Drawing on the feminist critique of techno-determinism in knowledge practices (Weber 2010), these performative workings of methods addressed the anxiety (of failure, pleasure) and constraints around negotiating the neutral view of research tools and centralized agency claims. As a result the release of surface energy and synergies in the final video blurs the boundaries between videography created for artistic, academic scholarship and everyday social media.

Conclusion

Thinking through the embodiment and affordances of everyday visual technologies such as smartphones and wearable cameras, the aim of this chapter was to present the practical and conceptual loops that were mobilized between a speculative research methodology and its material processes — shooting, editing, and the sharing of filmic knowledge — in this case, in the context of ordinary violence in Istanbul. The chapter focussed on a particular mediation of urban images animated by speculation as a research and theory-building strategy and raised several crucial questions. What does a speculative engagement with how violence appears in everyday environments ask of subjects and knowledge producers? What can be learned from everyday visual practices in terms of their embodied performativity within a material landscape of urban redevelopment? How might urban scholars and cultural and political geographers work with image, sound, and movement to explore spatiality and the creative body? And how might digital visual tools render imaginative dimensions of experience?

Where violence takes place and how it is remembered as it is visualized through participant-generated smartphone and wearable camera footage are not things that resist meaning, but rather problematize its disguise and raise wider questions about the epistemic burden of visualization in/for urban research within expanding systems of vision. The chapter first identifies this difference and then presents the evolution of the video *A Walk By the Shore*. The underlying determinism in visual representation dominating the design and validation of visual research from ethnographic film to big data processing and immersive environment design is a central theme of the discussion. I argued that under-problematizing social and material relations of visualization is embedded in the positivist desire for a "tunnel vision" between technological tools and subjects based on the assumption that "what is" or an already existing relationship between visual data and objects is to be rendered through the digital processes of visualization. This tunnel vision disembodies subjects and dematerializes the process, resulting in a relationship of distance and external identification with the visual environment. It also mutes the crucial discussion on how, when, and to whom accountability becomes an issue in terms of using and developing visual tools for knowledge practices.

The making of the research video presented an alternative case in which speculative coalitions were enabled by reshuffling and rethinking the role of everyday media in the way we remember/experience violence in the city. The methodology was speculative: expressive, responsible, sensitized, ways of doing and thinking the spatio-psychology of ordinary violence through images, footage, and conversations that resulted from a loosely formatted, dialogic research process. As a result, memories mediated through smartphones and body-cameras revealed that these visualization practices were never external to bodies but actually involved subjective, embodied complexes shaped by difference — mainly age and gender in the specific case of Istanbul (Duru 2018; 2019).

The past travels through technologies and practices to different bodies; it creates, enables and disables human and other (machine, platform, interface, code) subjectivities, to discover aesthetic possibilities, new meanings and new viewing experiences. The socio-materiality of methods (visual, digital) — what they do and their effects on politics of knowledge production — are vital in order to assess the visibility effects that result from the use of media and technologies in research and beyond. Re-routing visual practices around the speculative activates the politics of the story and counter-visibility and opens up an extended representational space to think through the whatness and whereness of research questions. The speculative is one strategy to examine the socio-materiality of media environments and what they do to us by incorporating the diversity of desires, counter-desires, and meanings attached to technologies. Speculative explorations of the relations between subjects and mediation affecting space and bodies are a key method of inquiring about the "human". Yet, this potential of the speculative further connects to why questions of style are under-theorized in visual cultural inquiry; this mainly has to do with an external view of technologies together with objectivism's persistent anxiety and inability to process pleasure and failure, as well as the pervasive assumptions around centralized agencies. Imaginative processes grounded in embodied knowledge about how bodies are enmeshed in wider socio-technical relationships offer gateways into critical inquiry by generating new questions and ways of engaging the gaps and possibilities for building collectivities, resistance and owning style, desire, and failure as invaluable effects of epistemic diversity.

References

Akrich, M. 1992. "The De-scription of Technological Objects." In *Shaping Technology/Building Society: Studies in Sociotechnical Change*, edited by Wiebe E. Bijker and John Law, 205–224. Cambridge, MA: MIT Press.

Allen, Louisa. 2011. "'Picture This': Using Photo-Methods in Research on Sexualities and Schooling'. *Qualitative Research* 11 (5): 487–504.

Ashburner, Xanthe. 2019. "Fabrizio Terranova (Dir.). Donna Haraway: Storytelling for Earthly Survival. Icarus Films, 2016." *Australian Feminist Studies* 34 (99): 128–130.

Auger, James. 2013. "Speculative Design: Crafting the Speculation." *Digital Creativity*, 24 (1): 11-35.

Axelrod, Robert. 1973. "Schema Theory: An Information Processing Model of Perception and Cognition." *American Political Science Review* 67 (4): 1248–1266.

Barad, K. 2007. *Meeting the Universe Halfway: Quantum Physics and the Entanglement of Matter and Meaning*. Durham, NC: Duke University Press.

Barad, K. 2014. "Diffracting Diffraction: Cutting Together-Apart." *Parallax* 20 (3): 168–87.

Barad, K. 2015. "TransMaterialities: Trans*/Matter/Realities and Queer Political Imaginings." *GLQ: A Journal of Lesbian and Gay Studies* 21 (2–3): 387–422.

Barthes, R. 1973. "Theory of the Text." In *Untying the Text: A Post-Structuralist Reader*, edited by Robert Young, translated by Ian McLeod, 31-48, Boston: Routledge, 1981.

Basu, Paul, and Sharon Macdonald. 2007. "Introduction: Experiments in Exhibition, Ethnography, Art, and Science." In *Exhibition Experiments*, edited by Sharon Macdonald and Paul Basu, 1–24. Oxford: Blackwell.

Bendon, Helen. 2005. "A Place for Ambiguity: Articulating Practice as Research." *Journal of Media Practice* 6 (3): 157–65.

Bennett, Jane. 2010. *Vibrant Matter: A Political Ecology of Things*. Durham: Duke University Press.

Bivens, Rena. 2018. "Exploiting a Dystopic Future to Unsettle Our Present-Day Thinking About Sexual Violence Prevention." *Ada New Media* (blog). 10 May 2018. https://adanewmedia.org/2018/05/issue13-bivens/.

Candy, Linda, and Ernest Edmonds. 2018. "Practice-Based Research in the Creative Arts: Foundations and Futures from the Front Line." *Leonardo* 51 (1): 63–69.

Chamaa, Sabine El. 2017. "Mind the Gap! Working Papers on Practice-Based Doctoral Research in the Creative Arts and Media." *Journal of Media Practice* 18 (1): 83–85.

Chapman, Owen, and Kim Sawchuk. 2012. "Creation: Intervention, Analysis and 'Family Resemblances'." *Canadian Journal of Communication* 37 (1): 5-26.

Davies, David. 2004. *Art As Performance*. Malden, MA: Blackwell.

Dunne, Anthony, and Fiona Raby. n.d. "SpeculativeEdu | Dunne & Raby: Designers Need to Recognise Their Limitations and Work with Them." Accessed February 8, 2021. https://speculativeedu.eu/interview-dunne-raby/.

Duru, Asli. 2018. "Wearable Cameras, in-Visible Breasts: Intimate Spatialities of Feminist Research with Wearable Camcorders in Istanbul." *Gender, Place & Culture* 25 (7): 939–54.

———. 2019. "'A Walk down the Shore': A Visual Geography of Ordinary Violence in Istanbul." *Environment and Planning D. Society and Space* 37 (6): 1064-1080.

———. 2020. "A Walk Down the Shore| Sahilde 1 Gezinti." *Cultural Geographies* 27 (1): 157–61.

Hackers of Resistance. 2018. "hord14r13z." *Ada: A Journal of Gender, New Media, and Technology* 13, https://dx.doi.org/10.5399/uo/ada.2018.13.6/.

Haraway, Donna. 2016. *Staying with the Trouble: Making Kin in the Chthulucene*. Durham, NC: Duke University Press.

Harding, Sandra. 1991. *Whose Science? Whose Knowledge?: Thinking from Women's Lives*. Ithaca: Cornell University Press.

Harper, Graeme. 2008. "Creative Writing: Words as Practice-Led Research." *Journal of Visual Art Practice* 7 (2): 161–71.

Hesse-Biber, Sharlene Nagy, and Amy J. Griffin. 2015. "Feminist Approaches to Multimethod and Mixed Methods Research." In *The Oxford Handbook of Multimethod and Mixed Methods Research Inquiry*, edited by Sharlene Nagy Hesse-Biber and R. Burke Johnson, 72–89, Oxford: Oxford University Press.

Imarisha, Walida. 2015. "Introduction." In *Octavia's Brood: Science Fiction Stories from Social Justice Movements*, edited by Adrienne Maree Brown and Walida Imarisha, Edinburgh: AK Press.

Johansson, Lotta. 2016. "Post-Qualitative Line of Flight and the Confabulative Conversation: A Methodological Ethnography." *International Journal of Qualitative Studies in Education* 29 (4): 445–66.

Konstantinidou, Eirini. 2017. "Practicing Theory: Mnemophrenia – A Film-Essay on the Future of Cinema and Artificial Memories." *Journal of Media Practice* 18 (1): 4–13.

Lather, Patti, and Elizabeth A. St Pierre. 2013. "Post-Qualitative Research." *International Journal of Qualitative Studies in Education* 26 (6): 629–33.

Law, John. 2004. *After Method: Mess in Social Science Research*. London: Routledge.

Lazzarato, Maurizio. 2019. *Videophilosophy: The Perception of Time in Post-Fordism*. Translated by Jay Hetrick. New York: Columbia University Press.

Leorke, Dale, and Christopher Wood. 2019. "'Alternative Ways of Being': Reimagining Locative Media Materiality through Speculative Fiction and Design." *Media Theory*, 3 (2): 63–102.

Lessig, Lawrence. 2009. "Remix: Making Art and Commerce Thrive in the Hybrid Economy." http://archive.org/details/LawrenceLessigRemix/.

Lyotard, J-F. 1984. *The Postmodern Condition (1979)*. Manchester: Manchester University Press.

Malinowska, Anna, and Toby Miller. 2017. "Sensitive Media." *Open Cultural Studies* 1 (1). https://doi.org/10.1515/culture-2017-0060.

Melitopoulos, Angela. 1999. "Passing Drama by Angela Melitopoulos, Excerpt, 4, 36." YouTube video, https://www.youtube.com/watch?v=1pAqy4Uoxjg.

Oxman, E. 2010. "Sensing the Image: Roland Barthes and the Affect of the Visual." *SubStance* 39 (2): 71–90.

Peake, Linda. 2016. "On Feminism and Feminist Allies in Knowledge Production in Urban Geography." *Urban Geography* 37 (6): 830–38.

Pelc, Noelany, Nadia Hasan, and Debra Mollen. 2020. "Feminist Storytelling: Representing the Stories of Diverse Women in Psychology." *Women & Therapy* 43 (1–2): 7–17.

Rose, Gillian. 2014. "On the Relation between 'Visual Research Methods' and Contemporary Visual Culture." *The Sociological Review* 62 (1): 24–46.

Rowlands, Mark. 2010. *The New Science of the Mind: From Extended Mind to Embodied Phenomenology.* London: MIT Press.

Skains, R. Lyle. 2018. "Creative Practice as Research: Discourse on Methodology." *Media Practice and Education* 19 (1): 82–97.

Toupin, Sophie. 2018. "Introduction: Radical Feminist Storytelling and Speculative Fiction: Creating New Worlds by Re-Imagining Hacking." *Ada New Media* (blog). 10 May 2018.

Weber, Jutta. 2010. "'Security' Architectures, New Ontologies and the Category of Gender. Contemporary Challenges in Feminist Technoscience Studies." In *GenderChange in Academia: Re-Mapping the Fields of Work, Knowledge, and Politics from a Gender Perspective,* edited by Birgit Riegraf, Brigitte Aulenbacher, Edit Kirsch-Auwärter, and Ursula Müller, 217–30. Wiesbaden: VS Verlag für Sozialwissenschaften.

About the author

Asli Duru is a feminist cultural-political geographer, mother, long-distance walker and Zen archer based in Berlin. She currently writes research-based fiction and non-fiction on home-spaces, ownership, bodies — imaginations, limits, boundaries. In a broader sense, she is interested in developing tools, design, and multimodal narrative using old and new media to activate formal and informal research and teaching practices that contribute to transdisciplinary knowledge on gender, difference and space/place relations.

8. Electronic Presence: Encounters as Sites of Emergent Publics in Mediated Cities

Zlatan Krajina

Abstract

Arguing that encounters among strangers are one source of publics in urban spaces, this chapter explores how publics change when encounters involve not merely humans but also what Kevin Robins terms "electronic presence". The chapter surveys debates about the public sphere in the urban context and more recent posthumanist interventions in understanding public cultures in mediated cities. The chapter discusses three case studies: people's daily encounters with outdoor advertising, highlighting the continuing relevance of power relations; social media posts about street encounters with xenophobic assaults, reminding us that visualizing can also mean distancing; and urban catastrophe relief campaigns in the absence of physical encounters, demonstrating that data about publics assumes a performative, rather than merely problem-solving role in urban living.

Keywords: encounter, public, city, urban, digital, media

Introduction

Public life stems from and defines urbanity. It is in cities that people can become aware of others and relate to shared issues. In negotiating terms of interaction, the urban world has embraced the digital, the mobile and particularly the visual. No longer recognized merely in expected sites like protest in public space, pub discussion, newspaper commentary, or nationwide broadcast, publics are also increasingly familiar in less explicit

---

Rose, G. (ed.), *Seeing the City Digitally. Processing Urban Space and Time.*
Taylor & Francis Group, 2022
DOI 10.5117/9789463727037_CH08

forms. These range from the murmuring of social media (Tierney 2013) to transient comments exchanged among locals and migrants in an ethnic shop (Valluvan 2016), the international transfer corridor (Back et al. 2012), the vandalized billboard (Krajina 2014), and surveillance footage from the traffic control room (Kitchin et al. 2017).

Prompting these mutations of how we understand publics is, in my view, and among other important processes, an ongoing transformation in the realm of visual culture. It is a slide from what Robins (1996) recognized as "electronic representation", that is, accomplished, coherent and authoritative narrations of city life provided by film and photography – made to be attended, viewed, and reflected upon – into radically dispersed and affective forms of video capture, "electronic presence". There is continued relevance of Robins' argument for understanding urban publics. It allows us to observe publics in moments when people's daily rounds in the city cross paths with diverse rivers of data, typically exemplified in public announcements, outdoor advertisements, news, self-presentation in social media, etc. Embedded, even at times imperceptible, in all kinds of private and public space (households, clothes, vehicles, façades), "electronic presence" invites an encounter with fragments of an inexistent totality of information. As opposed to accomplished narratives that intended to provide a space of withdrawal and analysis, which is "electronic representation", "electronic presence" tends to appear as suggestive but baffling, requiring one to develop skills of filtering, even at the expense of missing relevant information. Sensing the world as noise whilst being otherwise engaged – this is increasingly the dominant communicational setting for urban publics.

Issues of "electronic representation", like power, discourse, identity, and signification, have not lost significance in matters of "electronic presence". Its hardware, pertaining to processors and transmitters, appears diffused, yet its distribution, much like earlier pipelines and telephony cables, still concerns issues of uneven accessibility. And its content continues to reproduce community-building boundaries like codes of membership and behaviour (cf. Morley 2017). "Representation" and "presence" also share the basic site of interaction, the screen. Forms of screen sanctioned by late nineteenth century institutions, such as cinema, gallery and poster stand, exemplifying "representation", have now been joined by the more flexibly programmed and widely sited digital screens, characteristic of "presence". The latter kinds of screen are from the start designed to be "always ready for the next image [...] and constantly reconfigured by the hardware/software/user" (Hoelzl and Marie 2016, 372). Employed simultaneously in contemporary contexts, screens drawn from these different eras and logics come to create a rather demanding urban environment. Diverse as the modernist, spectacular storefront, the semi-visible surveillance monitor, and the networked, flickering mobile phone, the screen is now an assumed part of citizens' daily rounds in the city. It promises to supply citizens with information (location, description or space for interaction) and seeks to maintain connection (contact or control) with sites beyond the immediate space of interaction.

Ways in which this expanded space for interaction matters for urban publics remain ambivalent. Though diffuse and uncertain, "presence" continues to promise more opportunity for public voice as it involves in urban communication a wider than before range of actors and channels: not just news agencies, studios, or auteurs, but also citizens' posts and engineered algorithms. Technologies of simultaneous feedback and tele-presence (e.g. the smartphone, networked public screens) have lessened the importance of physical proximity in public interaction and participation. Thus, situations like bystander visual capture ("electronic presence") of police brutality against black people in US cities on handheld video devices (at least since VHS in the 1991 Rodney King beating, to use Robins' initial example, to smartphone-enabled video on social media in the 2020 George Floyd murder), have prompted wide-spread civic and then governmental action but perhaps only exceptionally. At the same time, the continued presence of such imagery in social media, which portrays streets as fearful, has become common knowledge for many black citizens and is likely to inspire further response.

In this chapter, I argue that landscapes of "electronic presence" encourage a longstanding (premodern) form of urban communication, the encounter, as a key source of (postmodern) publics. I observe the role that encounters, as passing but repeating and thus potentially transformative transactions among people, messages, and objects, can play in contemporary formations of publics. Devising an interdisciplinary approach as highlighted by "urban media studies" (Tosoni et al. 2019), I focus on encounters as urban entanglements of at once material, symbolic, and embodied aspects of public interaction. I argue that encounters, these slippery and seemingly insignificant situations, define imaginable scenarios for publics. If encounters function as "a crucial filter of social practice" in daily life (Amin 2008, 18), and if encounters leading to social movements "have dramatized the streets" (Merrifield 2012, 273), we need to explore what the perilous nature (brevity, semi-visibility) and multimodality (electronic, embodied) of encounters has to say about public interaction in mediated cities. I first trace encounters in debates about publics across communication, urban studies and politics and then move on to investigate more closely the terms on which encounters form and wither, in some typical sites where encounters crystalize as sources of publics. These include encounters with outdoor advertising, social media reports about encounters with xenophobic street assaults and public negotiation of urban catastrophe in the absence of physical encounters.

Acknowledging arguments for recognizing publics beyond congregative public space and broadcast media (Struppek 2006; Vuolteenaho et al. 2015), and post-humanist views that publics are constituted neither by human action nor technology like screens alone (Amin 2008), I focus on the largely ignored power-related conditions of encounters. I see encounters as always possibly multilocational and relatively uncertain situations of address-and-response among citizens and technologies, which rearticulate older issues of boundary and identity in new ways. My observation of publics through encounters appreciates their contingent nature, particularly in an Althusserian sense, whereby "nothing ever guarantees an encounter" (quoted in Merrifield 2012). This perspective also insists on keeping matters of difference in view, particularly given the challenging visual articulation of publics borne from encounters with "electronic presence".

Politics of visualization becomes an urgently relevant dimension of mediated urban publics as data gains ontological status in dominant evaluations of human practice and experience. Media regulators and industry alike tend to align "active" (engaged) media users only with a very specific (minor) element of media consumption practices, when users leave digital footprints of parts of their "activity" (e.g. "liking", sharing or commenting online), while offline commenting or intentional non-use is ignored. In the area of decision-making too, policies are thought legitimate if backed up by data, whether methodologically sound or not (e.g. Cronin 2010). This wider societal commitment to conceive everything, including publics, through data and visual presentation, does not also guarantee their political visibility and accountability. Mediated urban publics, so conceived, carry the potential to change, but this impulse, for better or worse, remains displaced or postponed. Further reasons for this situation lay in the constitution of the urban itself.

Urban cultures, public cultures

Conceivable encounters originate from a tension between order and improvisation. This duality is inherent to overall dynamics of city life, which becomes evident even from a preliminary, if narrowly European, historical reflection. The Greek polis, a political society in geographically detached city-states, was materialized in forms of agora (antecedent of squares) (Vallet 1985). The agora was open to all, but remained politically inaccessible to women, foreigners, and slaves (ibid.). Renaissance cities, particularly city ports like Amsterdam and Venice, relished socially diverse occupation of public spaces as sites of exchange and power (Braudel 1984); these cities also ascribed urbanity to a narrow definition of civility. The modernist conception of the public originated from industrial urban reconstruction, as exemplified by mid-nineteenth-century Paris. It heroized the historical relevance of the bourgeoisie for the fall of absolutism and planned the construction of spaces specifically intended for public interaction like streets and parks as part of a functional, and middle-class, city. Its famous boulevards, at the same time, marked the confines of acceptable public interaction; they were made wide enough to disable working-class protesters' barricades. The parallel rise of the nation-state and consumer society, with supporting infrastructures like unique measurement systems, money, transport, and telematics, distilled even further the modernist transformation of publics. Publics were relegated to displaced and spectacular sites such as those built of glass and illumination (e.g. shops and magazines) (Benjamin 1999), where public connection for assumed categories of citizens was couched in terms of glancing and gazing without confrontation, as is still familiar in much of daily social media use. The modernist city exemplified the "reformist liberal conception of public provision" of public space (streets, parks, leisure centres), a state-authorized spatialisation of the idea of publics (Bridge and Watson 2003, 371). By the mid-twentieth century, the positive value of mixed use of urban spaces and its visual culture (routine passing by as invisible civic safety net) became an urban planning policy issue, while typical forms of interaction done without contact, like "familiar stranger" and "abdication from responsibility", were found inevitable social facts in big city life. In fact, issues of migration and belonging in contemporary, transnational urban worlds have usefully disturbed the concept of "citizenship" as defined by state-issued criteria. The notion of "urban citizenship" (Lloyd 2013, 308) recognizes attachment to urban locale both by those with and without state citizenship.

The use of media – since at least late-nineteenth century inventions like the telegraph to the early twenty-first century successor, the networked mobile phone – further layered the dual (permitting and restricting) relevance of urbanity for publics. Media use, as argued above, disclosed a double sense of physical absence and virtual presence, as a legitimate factor in interaction. In the area of protest, we can still see people gathering in central urban spaces in cities like Soeul, Kyiv, New York, and Cairo. As they come together from across diverse private and public urban spaces, they now also require networked coordination in virtual spaces of social media to break through the clutter of personalized baffle and strategic, often commercially articulated, state-authorized communication (Willems 2019). People can now succeed in making their local civic claims known globally but they arguably only manage to operate at a slower pace than in a significantly different media ecology during the 1848 revolutions in central European cities (Therborn 2014). In other areas too media use has had ambivalent consequence for urban publics. Former world industrial hubs such as Newcastle have sought to transform into a service-based node in global trade, in which communication infrastructure has a key role. The city has also seen radically different modes of access to and use of digital communication in different, economically stratified, neighbourhoods (Crang et al. 2007). And those who seek to tackle uneven access to services, as in segregated neighbourhoods of Caracas, set up public spaces like community television, but less as a form of deliberation and more as subsidiary practical (routine) involvement in their overall struggle (Schiller 2020). Thus, the difficult interplay among spatial, societal, and technological conditions is not merely a backdrop but that which defines possible scenarios of public interaction, particularly its elementary situation, the encounter.

Encounters as publics: from congregation to speculation

If urban publics arise not from relations of kinship but those among strangers, encounters are the genome of cities. As Merrifield put it, "the urban was born from the encounter" among strangers who worked out forms of association (2012, 272). Encounters, these recursive episodes of noticing the existence and condition of the other, have helped set the scene for social change, such as the rise of the bourgeoisie (ibid.) and the emergence of the public sphere, a "realm" of critical reflection on public issues (Habermas 1991).

Habermas initially linked the possibility of a widely accessible public sphere with specific historical process (separation of civil society from the state), juridical categories (citizenship, representative democracy), spaces (cafés, streets), communicative form (intentional, rational, focused, directional and structured debate), and content (issues of "public interest"). It is now well-known that the public sphere so defined has never existed; neither has it ever been singular nor fully accessible. In fact, the rise of publics turned out having a lot to do with group boundaries, as a source of identities, and discussions of what counts as a public interest to whom, especially when it comes to groups invisible to mainstream debates (e.g. various minorities, seen pejoratively as part of undifferentiated crowds or masses). Public space has also been differentiated from public domain, to recognize that interaction may happen across various, including private, mobile, and virtual, kinds of space (Jensen 2020), particularly given that most city spaces are by now privately owned or managed (Bridge and Watson 2003, 372). Also opposing Habermas' early pessimistic disdain for market terms of interaction was a more inclusive argument contending that the market's niche system may provide visibility, or "representational" access, to those left outside the more conventional arena (Adut 2012, 243).

The urban, its mix of materiality, virtuality, and sociality, further unsettled the above re-definitions of the public. The early twentieth century urban scholarship of Simmel and Benjamin and later the work by Sennett, offered depth to the notion of public culture by looking beyond civil society as in Habermas. They focused on the dynamism of social life, one shaped not only by procedures of depersonalisation like the money economy, but also by new phenomena, the magic of the spectacular and the arguable loss of civic serendipity. To the extent that the city can be seen as "teatrum mundi", a stage for the performance of self-truth, these and especially post-war accounts calibrated urban publics as always-potentially, though rarely actualized as, progressive (Sennett 2003). As in earlier progressive periods (Mumford 1937), communication was considered a fundamental condition of urbanity, and public space instrumental in grasping social diversity. Moreover, contemporary interventions such as Mouffe's (2008) in recognizing insolvable agonism among viewpoints as the basis of democracy, have been exemplified by observations of urban practices like public art, which often expresses difference without seeking revolution or consent. Temporary occupations, performative interventions, and engaged billboard design serve as cases in point. Borne from converging lines of urban mobilities that bring entities into contact, encounters themselves are situations in which the emergent rather than predefined nature of identity can be observed (Ruddick 1996, 135). The encounter encapsulates the negotiation (acceptance, rejection) of discursive hails from surroundings. In turn, the public "comes into existence only – and always anew – in the moment of conflict and dispute [...] where it disappears, the public disappears together with it" (Marchart 2004, 16).

The increasing presence of large electronic display screens in public space, which, unlike printed billboards, can instantly switch to live broadcast or users' feedback, has invited a revival of interest in publics (see McQuire 2010 for a comprehensive critical history of such attempts). To enquire whether a planned use of large display screens might help "reinvigorating public space", McQuire et al (2013) commissioned a project where text messages describing personal values and places of origin by multi-ethnic participants in Melbourne, Australia and Incheon, South Korea were simultaneously projected on "mega screens" on main squares of these cities in 2008. The interaction led to "an experimental transnational public sphere", which depended on shared directions for use and functional technology with uninterrupted connection (328). Despite these limitations, producers, rightly, saw in participants' eager involvement an evocation of stoa, which was the stone-built "transitional" space between the Greek agora and the surroundings, where encounters among citizens and foreigners seeded the concept of cosmopolitanism as the positively valued urban experience of cultural difference (338).

Thus, there is a tendency to speculate whether mixed uses of big screens in highly frequented places to display adverts and public content might help a "rediscovery of the public sphere" in mediated cities (Struppek 2006, 173). The presence of such screens does not reduce issues of particularity and power, but merely articulates them in new ways. Selective logics of communication (the sharing of certain and not any or all codes of what defines dominant versions of common sense for different groups) continue to bracket any notion of universality when it comes to enjoying the benefits of open exchange (Morley 2000). For instance, in the realm of cultural publics, urban festivals only ever manage to attract certain rather than general audiences (Richards and Palmer 2010). Recalcitrant publics too may not always have their aims shared by all members, as seen in the coexistence of riots with protests (London in 2016, New York in 2020). Access itself is not merely about legally defined admittance to a place but also about recognition of tacit codes, which suggest who, among those formally admitted ("citizens"), is being spoken to, who is made more welcome and which type of behaviour is deemed more appropriate (Cresswell 1996). As exemplified in the 2016 London riots, a case of racist police shooting may bring back to awareness issues of systemic racial inequality that belong to unquestioned aspects of daily life, such as encountering, within accessible public spaces of streets, inaccessible goods, seen in store fronts during daily rounds. If accessible public spaces have had very different meanings for different groups, pleasurable promenading for men and whites and unquestioned insecurity for women and non-whites (Stevenson 2003; Towns 2020), there is no reason to assume that boundaries will diminish in spaces overlaid by communication technologies.

While a rising scholarly interest in affective dimensions of the urban and its posthuman (i.e. not-merely-human) configuration, persuasively decentres human action from working definitions of publics, it does so at the cost of appreciating continued issues of practical involvement, struggle, and disagreement. These tenets, as the remainder of this chapter will demonstrate, remain central to public culture. The posthuman observes dense sites with people moving and messages flashing positively as a "pre-cognitive template for civic and political behaviour [...] a distinctive sense of urban collective culture and civic affirmation" (Amin 2008, 5). The posthuman builds from an earlier, phenomenological (human) perspective, which discussed "place choreography" (Seamon 1979) to shift emphasis from a purely physical, measured, controlled, planned, projected and cognitive (Cartesian) understanding of space to unspeakable, embodied routines that create places known only intimately (genius loci). In its urban analysis, the posthuman further affirms the relevance of feeling to argue that each place in fact possesses a shared ambiance not managed directly or solely by humans. For instance, Sassen recognized such impressions (without engaging with the posthuman approach explicitly), as "an elusive urban capability [...] city's speech" (2014, 38). She referred to performative capacities of any typical site like a street corner, where walking, traffic, street vendors, lights (ibid.) together create a sense of heterogeneous (multiscalar, multidimensional) pulsation of activity. This perspective is an invitation to "reimagining the urban", in which the "encounter, and the reaction to it, is a formative element" (Amin and Thrift 2002, 30).

Though rightly recognizing encounters at the nexus of experience, space, and technology, as thus far underexplored sources of publics, the generalizing rhetoric of the posthuman perspective fails to acknowledge how people work through specific encounters. As my following three case studies will show, looking deeper into the dynamics of encounters, in rare opportunities to freeze-frame aspects of encounters for study, uncovers an unceasing struggle for meaning. First, my case of people's daily encounters with advertising screens, which address people as captive and people respond creatively to reject the address and maintain a routine movement, will remind us that power relations between institutions and citizens continue to matter during encounters. My second case, pertaining to pacifying networked responses to conflictual encounters in the street will highlight that screening as a mode of visualizing encounters serves not only disclosing but also distancing. Lastly, my observation of reliance on digital technologies in urban catastrophe relief will confirm that affect is relevant but not an exhaustive element: people still seek structure like narratives and maintain difference in their encounters with the urban to continue making it meaningfully habitable.

The everyday publics of outdoor advertising

Mediated cities are not just spaces where a lot of communication happens in different ways. These are spaces where communication is *encouraged*. Here, communication serves a highly performative function underpinned by a post-industrial logic according to which being constantly connected, alert and responsive is a necessary condition of success. Positioned strategically within everyday spaces to strike conversation with passers-by, the presence of advertisements for products and services, vocalizing currently dominant lifestyles and values, steadily grows. Given the well-known impossibility of opening the "black box" of consumer behaviour, outdoor advertising continually seeks new ways, without abandoning previous attempts, to arrest attention and produce "captive" audiences (Gitlin 2001, 68). The assumption is that the more and bigger screens there are in public space, the less chance citizens would have for escape. Ways in which publics emerge when people negotiate this situation has, perhaps counter-intuitively, a lot to do with household television cultures and relational constitution of privacy.

The availability of remote control in the family dynamics of household television drove the creation of "flow", as the central organizing logic of broadcast (pre-platform) era of television, based on frequent advertisement interruptions intended to keep viewers from flicking channels (Williams 1976). In public spaces, it is the absence of the remote control for screens that advertisers seek to profit from, without ever being able to achieve a complete "'privatization' of the space via the screen" (McCarthy 2005, 124). As I describe below, a typical large display screen in a residential street can be fixed against a brutal structure of an electricity generator, seen also from inside the surrounding apartments, and showing brief stills of news and ads to those passing by. These messages are usually designed to stand out from the street clutter (busy traffic, flaneuring, blasé walking, nearby construction work etc.) but can only invite distracted viewing, familiar in the household setting. And while household space may be sealed off by walls, the ads seen in the street are likely to reappear on television and mobile phone screens, both continuing to compete for user's attention.

It is at this dynamic intersection between the private and the public that people's encounters with outdoor advertising as a site of negotiation is to be found. In her research on connections between the urban and outdoor advertising, Cronin (2010) found that, observed together across various locations, billboards seem to visualize a "metabolism" of the city. They not only change messages to match rhythmic patterning of passers-by activity (e.g. promoting food deals to those going and evening events to those returning from work). Billboards also visualize a particular set of assumptions, held by advertisers, in terms of where people move and how they communicate during movement. Poster positions, layouts and contents together present a positive valorisation of perpetual mobility and interest in new things (2010). This is also the case in the so-called Silicon (Old Street) Roundabout in London, where the presence of advertising screens supports the nation's vision of a successful post-industrial city. Showcasing latest improvements in display tech and graphic design, outdoor advertising, there and elsewhere, becomes a visual megaphone within the street. Advertisers' preference for "bold", "striking" and "simple" design (Cronin 2010) is meant to have the message shout, among other sources of public noise, "look over here". Thus, the overused and derelict passage during the day becomes a luminous advertising scenography seen glowing against the darkened background in the night when the site takes on a visual cover for its continued physical decay. By analogy to the theatre stage, where darkening the auditorium serves to focus the attention of audiences, advertising screens offer a temporary gate to another, more colourful, reality, in which passers-by could, for a moment, imaginatively escape the grim roundabout. Ironically, the screen's visuals are easily noticed replicated in countless other, less sophisticated, versions, as posters on public transport and waiting rooms or ads in magazines and social media feeds.

My ethnographic study in this location (quoted below from published and unpublished materials pertaining to Krajina, 2014) found that people respond to repeated invitations to communication from advertising screens creatively. They negotiate the terms of situation (encountering the screen as presence to avoid or to pretend observing to reject the gaze of another passer-by), spatial arrangement (reading screen's illumination as gratifying symbolisation of a big city), issues of belonging and loneliness (interpreting constant promotion of new products as uncanny changes in the appearance of the environment) and consumption as a form of participation in society (evaluating graphic layout of advertisements whilst waiting to continue moving).

In walking diaries (audio recordings of thoughts had during walking and looking around) and subsequent interviews, my participants documented reviewing the sensorial arrangement of space on encountering an advertising screen. As one respondent described, with screens, "it's not like [...] when you pass the people in the street and you move each other's way", the encountered image is simply, "there"; it "sort of comes into my sphere [...] pops in-pops out". Screens do not "reciprocate" attention (Gitlin 2001, 20). Their presence mattered for the perception of space so that witnessing an image meant witnessing something happening in the street. For another respondent, who passed by a phone box and saw a realistic image of a child lying on the pavement after being hit by a runaway car (as part of an anti-speeding campaign), experiencing actual traffic was shot through with a sense of anxiety about a likely dramatic (unheimlich) event that threatens to disturb the seeming orderliness of the street. Conversely, as another respondent noted, a smile on a "Coke advert" can be read as "lift(ing)" the mood in the space that otherwise feels like a non-place (Krajina 2014, 77). Such issues of character, far more diffuse than the public speech act ("communicative action") considered by Habermas (1996) as one essential condition for the public sphere, matter too for publics as they inform a sense of (non)belonging that can become a point of political struggle.

In an environment, such as the mediated city, characterized, in busy sites, by simultaneous sensations vibrating in various, and not entirely disharmonious, directions, rhythms and intensities, communication is often about moments, frustration, and rejection rather than continued involvement, pleasure and commitment. My respondents sought ways to routinely evade the attempts of advertising to initiate communication, by looking away during movement, thinking "I'm gone now [...] Bye!" Those who stood waiting for someone looked towards the screen and pretended to read it, to alleviate the discomfort of the gaze of another passer-by. In a city where "the eyes need something to do, rather than merely see", as one respondent put it, moving on means having to accept to communicate with something else, such as smelling nearby kebabs rather than reading the visuals of a screen, or carrying on a conversation in one's mind about an advert's social, usually sexist, representation, after the screen has left one's sightline. Publicity concerning consumption becomes a topic for both private consideration and social participation. Encountering a "Honolulu for £1" advert, can provoke negotiation of its public address, or as one respondent said while passing it by, "I don't need that. I want to need Honolulu for £1 when I want to go to Honolulu!" (Krajina 2014, 88). Over time, she developed a routine of not looking at any public announcements, even at the cost of missing potentially useful information. For many, screening out publicity indiscriminately becomes a way of participating in the urban society, in which, "I've been faced with information that's irrelevant to me and I feel like I wasted time" (96).

Moreover, encountering the screen may invite passers-by to reflect on the identity of place through autobiography. As Featherstone put it, "the mobility of the gaze and the swings between immersion and detachment help to develop an attitude in which the urban landscape itself becomes perceived as fragmented yet allegoric" (1998, 915). The presence of the luminous screen can suggest that the neighbourhood in question has been recognized by wider flows of trade as valuable and settling there can be a testimony of personal success. For such respondents, typically newcomers, "billboards nearly as big as buildings [...] for an ad!" were a source of fascination which "you see [...] on TV, in Japan, New York [...] with a lot of communication and technology" (Krajina 2014, 95). For locals too, encountering the screen occasions reflection on the lifecycle of the neighbourhood. Those who walk past the same interfaces every day found the routine ignoring of the screens, possible after having tactically memorized their locations, layouts and range of typical content, as a source of attachment (a sense of place). The experience was similar to embodied and pre-conscious sensing of continued presence of other street inventory like benches. Unlike benches, the appearance of screens continues to change from a hidden control station in order to keep their incidental users attentive. Ensuing encounters with screens are thus about people's struggle of making – and invisibly maintaining – a public site one's own, that is, giving the public screen private meanings and episodically ignoring the screens.

In that context, encountering a new screen can trigger an observation of wider change of the neighbourhood. A lifelong local only made sense of the then new screen in terms of parallel volumes of new bars and restaurants springing up along with new real estate developments, where "none of that used to be" (Krajina 2016, 53). Older buildings like the "most amazing Victorian tea houses" in the vicinity were "knocked down", making him feel, whilst passing by the new screen, that the direction of change of his neighbourhood was such that "part of our history (is) just gone forever". Some heritage sites are preserved, by way of repurposing, but "you start to lose track of what things used to be there" (52). Such reflective moments of encounter, usually passed undocumented, are significant in challenging the ideological premise of gentrification projects that space is ahistorical and sweeping urban change is unsurpassable. Thus, what was "difficult not to notice" for another respondent was when, on a back-lit printed poster showing a famous singer someone had "cut like a slip through her eye and poured this like really bright green paint [...] coming out of her eye! (laughter) This was so cool!" It was unusual, in a slightly cathartic way. The changing appearance of electronic screens covering portions of solid urban surfaces and managed from hidden control rooms, invalidates civic attempts at material intervention in printed posters like drawing moustaches, writing comments, or sticking chewing-gums. Even if passers-by "leave no trace" (Ingold 2004, 329) of their daily rounds, their encounters with digital interfaces are never monologue.

Silent passing encounters are a mobile space of private interaction with public issues. Encounters remain invisible to procedures of datafication, the limitations of which become clearer in moments of conflict.

The publics of conflict in urban space mediated by social media and smartphones

If Wirth's classic postulate still holds, even at times of anti-pandemic measures demanding physical distance, that public urban space is defined by "frequent close physical contact, coupled with great social distance" (1938), electronic contact too matters for encounters. Communication about impressions had whilst moving through the city allows one to "possess its traces" (Gordon 2010, 3) beyond moments of passing through a city's physical spaces, on which I focused in the previous case. The novel urban phenomenon resulting from electronic capture of physical encounter, the "digital street" (Lane 2016), is space occupied both physically and virtually, through embodied use of the networked mobile phone. Mobile phones have tapped into the historical relevance of mobility for the constitution of the urban; they fuse virtual with physical mobility, providing certain people access to issues of public domain virtually whilst being on the move physically (Jensen 2020). These new, "mobile agoras" may be "stretched across time and space" but they are perhaps best understood in relation to parallel investment in physical urban highways, which may become one substitute for the central square in polycentric megacities (Jensen 2020). Contrary to their promises of inclusivity, physical and virtual spaces in which people congregate through movement reproduce issues like "who gets asked 'where do you come from?', what kinds of bodies and identities are allowed to circulate" (Lloyd 2013, 312; cf. Towns 2020). Indeed, when strangers go online to comment on their earlier physical encounters with other strangers, both material and virtual spaces turn out to be contentious and deeply interrelated.

As much as sights of routine street life reproduce a sense of normality (as the previous case study suggested), encounters can in any moment unlock underlying issues. Civic assault at those deemed dangerous strangers may momentarily elucidate unresolved issues that originate elsewhere (e.g. national policy of attracting cheaper migrant labour while disregarding the local workforce), bringing different groups of disadvantaged strangers into proximity. Diverse definitions of common sense, concerning what is accepted to do (ignore, shout) on encountering which kinds of others (white, non-white) in which contrasting urban universes, may lead to physical confrontation but also virtual conciliation (and vice-versa).

As I wrote in more detail elsewhere (Krajina, 2019), following the 2016 victory of the Leave vote in the Brexit referendum, social media circulated images of posts by witnesses of assaults (someone passing hateful remarks or physical attacks) which they had witnessed against non-white and white non-UK citizens (particularly Eastern Europeans) in public spaces in UK cities. Encounters were the principal sites of assault and bystander reaction and thus they became the key social situation quoted nationally to discuss the state of multiculturalism (there were more than half of the national average number of reports of hate crimes more than in the same week a year earlier). If "images" of others circulating the media "serve to prefigure unplanned encounters" in public space (Ruddick 1996, 139), it was mainly the new, eastern EU migrants, who were targeted by stereotyping as stealing jobs and welfare and were scapegoated in unresolved class and racial antagonisms in the country. Mobility of data via smartphones intersected with the transience of street encounters and the posts about assaults mostly adopted a headlines form, sometimes also assessing the observed situation (e.g. "we must tackle racism"). The smartphone screen functioned as a way of both disclosing the crime virtually and distancing the bystander physically from the troubled scenes. As Robins put it, reflecting on the uses of "electronic presence" in urban living, "individuals strive to protect themselves against the lurking and encroaching threats of the city" (1996, 139). Using portable recording devices, "they aspire to keep dangers at a distance" (139). Virtual witnesses grew in numbers by way of sharing such information on social media and thus extended the scale of public interaction (typically a local sidewalk or bus stop) to the scale of national and European media, which picked up on these events as a resonance of Brexit. Social media posts about street assaults, examined in this case, gave inter-personal encounters publicity but rarely engaged in considering the wider crisis. Publics thus created were as it were half-complete: vocal but without demands or goals; spatial but dispersed; themed but unfocused. As I show in the next and final case, agonism continues to characterize publics even in the absence of physical encounter.

Infrastructural scenographies of data publics

While the previous case highlighted the agonistic character of urban publics, this final case will explore its possibilities in times of limited physical contact. The truly global urban crisis caused by the Covid-19 pandemic made encounters a source of anxiety concerning pressures to track passers-by who may be asymptomatic yet infected by the unknown virus and reconstructing the whereabouts of those confirmed infected. This crisis reminded us of the classic assumption in urban governance, as old as cities, that the more we (think we) know about life in the city, the safer we feel as its inhabitants.

Ever since projection technologies of photography and film in the late-nineteenth century were used to chart urban territories, out of necessity (spaces were too large for the human eye) and artistic reflection (spaces were profoundly new), things visible and visualized were represented as legible and safe, while dark spaces implied danger (e.g. in film noir). States have consistently used urban data (postal addresses, video, GPS data, etc.) to seek a reliable monitoring, learning and prediction of patterns of urban living. This quest has made gathering and processing of information an indispensable ingredient of urban governance. Moreover, post-industrial cities have made services, including communication, a primary source of income, investing in projects inspired by visions of a datafied, luminous, clean, and egalitarian, though faraway, future. These imaginaries set "vision and light against the forces of darkness" (Robins 1996, 130). They depict preferred urban spaces as flashing landscapes crisscrossed tirelessly by constant feedback and cutting-edge technology. Though unevenly distributed, installed within existing old, often deficient structures, and of high maintenance, technologies of digital command for functions like transport, are represented through images of "uninterrupted mobility" (Rose 2017, 9). Thus, paralleling the permanent incompleteness of "smart city" technologies is the branding of a flowing, networked city as a good city, even though it is a very specific, usually utopian/futuristic and "depoliticize(d)" (Shelton 2017, 3) an understanding of urbanity.

Modernist institutions like statistical offices, dispatch services and newsrooms have seen a dispersal of authority in information collection, which now also involves automated (corporate) and civic input. In this environment, publics are not only hailed institutionally, through spatial design or urban media, but are also seamlessly constructed through transferable data, harvested from various information terminals. A sensor responsive to momentary traffic may inform semaphore signalling but also crime anticipation, advertisement positionings and long-term development of neighbourhoods. Translated into urban policy, data helps construct multi-scalar but also multi-modal urban publics. Automated, yet elastic (allowing real-time change of parameter), and always partial (running on "algorithm bias" such as in racial profiling of offenders and their neighbourhoods in visualizing "future crime" [Erickson 2020, 232]), data publics are a performative post ante document of urban living.

As opposed to how heroic architecture (squares, monuments) imagined publics as subjects hailed to gather as members of a nation, and how modernist urban media institutions such as newsrooms performed "enculturation into big city living" and "journalistic place-making" (Rodgers 2020, 68-9), data publics do not necessarily belong to a localized or accomplished category like worker, veteran, believer. As very diverse discourses permeate citizens in actual space too, so in virtual space "clusters of networked publics" also form "multinucleated patterns" based on complex belongings in terms of class, race, ethnicity, taste, cultural capital, etc. (Tierney 2013, 66). Serving different projects (e.g. political or brand awareness campaigns), it is the performativity of data publics in the situational scenography of specific urban space that becomes most evident, particularly at moments of abrupt change in the datafied inhabitancy of urban space, such as power outage or natural disaster.

When a 5.4 magnitude earthquake struck Zagreb during the Covid-19 pandemic lockdown in Croatia in March 2020 (also temporarily disabling buildings like hospitals and the parliament), the very next day the city government launched an internet site for citizens to supply information (description and photos) about damage on their property, which was transcribed onto the official city's geodetic map and used to coordinate specialists' physical inspections. They were given a custom-made mobile phone app for reporting directly into the centralized register that could calm the anxious public by demonstrating consolidated information on damage quickly (left image on Figure 1). While data collection required constant eye-level movement between virtual (user interface) and physical (buildings) spaces, data visualisation took a birds-eye view (a map), which has always served urban authorities to profess a sense of control. And while this performative use of data was effective in persuading the public in authorities' successful command of the situation, a more functional use of data, for administering physical repairs, especially financial aid, took a separate course, hampering the speed of recovery.

Some fifteen per cent of citizens, possibly due to deficient computer skills, filed paper forms that required digitalisation. Afterwards, all digital and digitalized damage reports had to be printed back for appropriate legal processing (right image on Figure 1). Furthermore, due to Zagreb's capital status, relevant state institutions requested a separate documenting of damage as well as transferring the local authorities' data manually into the state's different software (which took another several months). A year on,

Figure 8.1. Damage data. Photos by author.

city officers can access comprehensive data for each location (addresses, information on owners, photos, descriptions) and simultaneously keep an eye on the overall statistics of processed claims (left image on Figure 1). Major repairs still haven't started.

Soon after the earthquake, a network of commercially run public display screens in the most badly damaged city centre, which had featured ads and public service information, switched to civic messages of support (similar to broadcast television's "programme interruption"). The glossy, high definition moving image screens flickered with notifications to absent passers-by saying things like: "Beloved city, hang in there" and "We'll come out stronger out of this". Discursively homogenic (vocalizing distanced care and positive tone, rather than local fear or frustration), regulated as brief, in full sentences and without specifying senders or receivers of messages (unlike plain paper information about provision of help for the elderly sellotaped on entrances to buildings), these digital gestures of morale dominated the screens as commercial activity had significantly gone down. Invisible citizens were seen speaking to absent inhabitants, thousands of whom had fled due to structural damage and fear. The messages on these clear and luminous graphics, devoid of dust, ambiguity, or error, contrasted with the surrounding postapocalyptic scenery of debris, traffic lights changing without traffic to regulate and businesses closed without notice of reopening (Figure 2). Cutting-edge visuals, used for marketing commodities, now spoke of loss and uncertainty. As argued earlier, the post-cinematic visualisation of publics is uncommitted but responsive.

This strange, advertisement-like visualisation of abrupt collective trauma was also a curious gesture of reassurance: the luminous screens signalled that electricity and computing, vital urban infrastructures, were still operational. The persistent gloss also evoked glimpses of vitality, emblematic of modern cities since floodlighting and neon. Spilt on the surrounding concrete surfaces, which turned out to be more fragile than images they had carried, the warm glow from the screen was impersonal, but protective. Life in public space had halted, where it had seemed imminent, but the screens went on screening in a nonchalant manner, the same in which they normally get covered by those passing by, without anyone requesting clear viewing space. The screens kept speaking when there was nobody to listen. Yet they strangely contoured a moment of shared fate, one that would have not existed without being represented, and they articulated a consensus, felt elsewhere, in the media, about the need for support. In the night-time, these desolate screening sites seemed not merely escapist, as in the Old Street case, but somehow celestial.

Figure 8.2. Messages. Photos by author.

Information about this screening action was to be sought elsewhere, online, where the company running the projections set up a web platform to collect "Messages for Zagreb" (Figure 2). Displaced too were passers-by's responses as the campaign ran during the lockdown so people could only really see these messages on photos of the displays shared on social media. Several thousand likes and several dozen comments on the company's Facebook page suggested an overall recognition of relevance but the comments also indicated agonic negotiation of electronic presence of these anonymous supportive others in deserted streets. Some locals reflected on the one-dimensional character of the sentimental/melodramatic announcements as "silly" and inappropriate, given the state of emergency and delayed aid, others criticized the city government's longstanding neglect of historic architecture, and others the invisibility of suburbs in such campaigns.

Thus, the screens displayed a sentiment of endurance and support without the latter being given and taken. This intervention in physical public space attracted displaced, online public discussion concerning the actual disaster.

From a posthuman perspective, one "not centred on the human experience [...] without dismissing the concept", encountering a screened civic voice was about meeting "impersonal forces" woven into "the worldly textures and trajectories of that life" (McCormack 2017, 2, 7). The "post-phenomenological" lifeworld emerges from "circumstantial worlding of forces excessive of the subject" (2). This world of "situated multiplicity" relates to everything composing the horizon of the situation, "bodies, mass and matter" (Amin 2008, 8, 11). Multiplicity surpasses one's individual will or cognition and is about co-joined and emergent "circuits of flow and association that are not reducible to the urban" (8, 11, 6).

This theorized multiplicity usefully describes publics as multimodal but has less to offer in our efforts to account for how publics negotiate issues. The posthuman argues that "interaction is not a sufficient condition of public culture", which is rather considered to result from a "swirl of surplus" of things, signals and bodies (Amin 2008, 8, 11). As illustrated by typical images of bustling markets, parks, libraries, central city crossroads or parkour sites like rooftops, this "surplus" itself likely fosters an air of conviviality and care (ibid.). The city overall is seen as "an ordering of uncertainty" (Amin and Thrift 2002, 77). Missing from this view is a consideration of responsibility for the conditions (spaces and institutions) for public interaction because the outcomes of public interaction continue to depend on the humans engaged (McQuire et al. arrived at a similar conclusion in commissioning public screenings, 2013, 336). Reminding us, correctly, that the centuries old urbanist desire to engineer civic interaction through spatial design failed because it misunderstood conviviality for disorder, the posthuman links the possibility of public culture to moments of "non-hierarchical relations" when places escape the command of pre-existing vision, without rejecting issues of power entirely (Amin 2008, 11).

To return to Zagreb's campaign, the question arises, whether a sense of such self-governing energy ("force" [Amin 2008, 11]) is typically shared across different groups, particularly those citizens who usually lack opportunity to voice their views widely and whether it is helpful to observe *any* urban place outside, even implicit, structures (narratives) and hierarchies (power relations). Mediated urban publics do emerge from an uneasy fit among material infrastructure, social reality and data, crossing paths often without obvious logical connection or narration (cf. Robins 1996). As we also saw in cases of London Old Street advertising and Brexit assaults, passers-by made seemingly chaotic encounters meaningful *as* narratives. They routinely translated electronic presence into representation as they conversed with perceived and hierarchized social realities such as not belonging in gentrified space, xenophobia, or delayed disaster relief. Digital systems in segments of urban life draw forever uneven partners (institutions and pedestrians) into contingent spaces of control and response that may lead to affective "character" but one which beneath the surface-level observation remains agonistic and partial.

Conclusion

Encounters are an underexplored urban epistemology of public culture. From this perspective, the city is neither a thing nor merely a process but a "constant encounter": an outcome of "interactions, connections", "a cum" (Nancy 2009). To understand how publics emerge from encounters requires us to recognize that encounters are not merely a matter of "chance", as their popular definition suggests, but embody or conceal diverse forms of social patterning. There is a cultural geography of encounters that concerns how conceivable encounters are conditioned by patterns of mobility (gendered, classed and racialized coming-into-touch), that is, who is likely to meet whom, where, how and with which consequences. Multivalence and undecidability of contexts in which I observe the formation of publics did not make issues of power, such as curation and distribution of technologies for publics like screens, peripheral. Uneven positions from which authorities and citizens are drawn to speak to each other via visualizing digital devices are likely to be reproduced rather than transformed. Whether the omnipresent mobile phone or the elevated public display, screens pierce messy physical and embodied urban realities with evocations of authority through the medium of the visual: digits are made to seem brighter, clearer, and more reliable than any embodied or spatial aspect of the urban can ever be.

The rise of the public sphere historically depended on a bracketing of the individual and the state, while modern urban society was made possible through the private gaze exchanged in movement and without confrontation. Intersections of the visual/digital ("electronic presence") with walking, glimpsed in the first; conflict, observed in the second; and catastrophe, considered in my third case study, intensify the above tendency. Mediated urban publics appreciate confrontation through its radical displacement and postponement. So arranged, publics articulate through the language of the visual: in encounters with advertising as display and invisibility, in social media posts about assaults as disclosing and distancing, and in responses to catastrophe as both reassuring and insufficient.

Mediated urban publics thus observed were:

a) integrative of different modalities of action (embodied, material, and algorithmic), forms (e.g. news feeds read alongside advertisements), scales of articulation (local and beyond), and temporality (instant feedback but postponed consequence)

b) diffused, that is, not necessarily oriented toward political change or aim, nor being consciously assembled or reflective.

Encounters remain difficult to arrest for analysis and thus their sediment of alterity in principle remains unknown and, luckily, uncontrolled. Having unknown beginnings and ends, yet unceasing and multiple (always possibly implicated elsewhere via networked screens, where they may have other relevant consequences), encounters make it impossible to map out all public impulses and their possible directions. Because they are borne out of encounters, online and offline, publics may emerge where and when they are least expected.

References

Adut, Ari. 2012. "A Theory of the Public Sphere." *Sociological Theory* 30 (4): 238-262.

Amin, Ash. 2008. "Collective Culture and Urban Public Space." *City* 12 (1): 5-24.

Amin, Ash and Thrift, Nigel. 2002. *Cities: Reimagining the Urban*. Cambridge: Polity.

Back, Les, et al. 2012. "New Hierarchies of Belonging." *European Journal of Cultural Studies*. 15 (2): 139–154.

Benjamin, Walter. 1999. *The Arcades Project*. Cambridge, MA: Harvard University Press.

Braudel, Fernand. 1984. *The Perspective Of The World*. New York: Harper & Rowe.
Bridge, Gary and Watson, Sophie. 2003. "City Publics." In *A Companion to the City*, edited by Gary Bridge and Sophie Watson, 369–379. Oxford: Blackwell.
Crang, Michael, et al. 2007. "Technology, Time-Space, and the Remediation of Neighbourhood Life." *Environment and Planning A* 39: 2405-2422.
Cresswell, Tim. 1996. *In Place/Out of Place: Geography, Ideology, and Transgression*. Minneapolis: University of Minnesota Press.
Cronin, Anne M. 2010. *Advertising, Commercial Spaces and the Urban*. Basingstoke: Palgrave Macmillan.
Erickson, Kristin. 2020. "The Promises and Pitfalls of Cyber Urbanism: Governance and Participation." In *The Routledge Companion to Urban Media and Communication*, edited by Zlatan Krajina and Deborah Stevenson, 225-235. London and New York: Routledge.
Featherstone, Mike. 1998. "The 'Flâneur', the City and Virtual Public Life." *Urban Studies* 35 (5/6): 909-925.
Gitlin, Todd. 2001. *Media Unlimited: How the Torrent of Images and Sounds Overwhelms Our Lives*. Metropolitan Books: New York.
Gordon, Eric. 2010. *The Urban Spectator: American Concept-Cities from Kodak to Google*. Lebanon, CT: University Press of New England.
Habermas, Jürgen. 1991. *Structural Transformation of the Public Sphere: An Inquiry into a Category of Bourgeois Society*. Cambridge, MA: MIT Press.
Habermas, Jürgen. 1996. *Between Facts and Norms: Contributions to a Discourse Theory Of Law And Democracy*. Cambridge MA: MIT Press.
Hoelzl, Ingrid and Marie, Rémi. 2016. "Brave New City: The Image of the Urban Data-Space." *Visual Communication* 15 (3): 371-391.
Ingold, Tim. 2004. "Culture on the Ground: World Perceived through the Feet." *Journal of Material Culture* 9 (3): 315–340.
Jensen, Ole. 2020. "On the Move: On Mobile Agoras, Networked Selves, and the Contemporary City." In *The Routledge Companion to Urban Media and Communication,* edited by Zlatan Krajina and Deborah Stevenson, 96-106. London and New York: Routledge.
Kitchin, Rob, et al. 2017. "Urban Informatics, Governmentality and the Logics of Urban Control: The Programmable City Working Paper 25" http://progcity.maynoothuniversity.ie
Krajina, Zlatan. 2014. *Negotiating the Mediated City: Everyday Encounters with Public Screens*. London and New York: Routledge.
Krajina, Zlatan. 2016. "The Alternative Urbanism of Psychogeography in the Mediated City." In *Radical Space: Exploring Politics and Practice*, edited by Debra Shaw and Maggie Humm, 39-63. London: Rowman & Littlefield.

Krajina, Zlatan. 2019. “Understanding Encounters for Urban Media Studies: Civic Intercourse, Screen Technologies, and Cultural Difference.” *International Journal of Communication* 13: 5351–5369.

Lane, Joshua. 2016. “The Digital Street: An Ethnographic Study of Networked Street Life In Harlem.” *American Behavioural Scientist* 60 (1): 43-58.

Lloyd, Justine. 2013. “Trouble Comes to Me: The Mediated Place of the Urban Citizen.” *Space and Culture* 16 (3): 306-322.

Marchart, Oliver. 2004. “Politics and Artistic Practice: On the Aesthetics of the Public Sphere.” *Frakcija* 33/34: 14–19.

McCarthy, Anna. 2001. *Ambient Television: Visual Culture and Public Space*. Durham, NC: Duke University Press.

McCormack, Derek. 2017. “The Circumstances of Post-Phenomenological Life Worlds.” *Transactions of the institute of British geographers* 42: 2-13.

McQuire, Scott. 2010. “Rethinking Media Events: Large Screens, Public Space Broadcasting and Beyond.” *New Media and Society* 12 (4): 567-582.

McQuire Scott et al. 2013. “Mega Screens for Mega Cities.” *Theory, Culture and Society* 30 (7/8): 325-341.

Merrifield, Andy. 2012. “The Politics of the Encounter and the Urbanization of the World.” *City* 16 (3): 269-283.

Morley, David. 2000. *Home Territories: Media, Mobility, and Identity*. London: Routledge.

Morley, David. 2017. *Communications and Mobility: The Migrant, the Mobile Phone, and the Container Box*. London: Wiley-Blackwell.

Mouffe, Chantal. 2008. “Art and Democracy: Art as an Agonistic Intervention in Public Space.” In *Art as Public Issue: How Art and Its Institutions Reinvent the Public Dimension* 14: 6-13. Rotterdam: NAi Publishers.

Mumford, Lewis. 1937. “What Is A City?” *Architectural Record* 82: 52–68.

Nancy, Jean-Luc. 2009. “Un art de la ville.” *Caliban: French journal of English studies* 25: https://journals.openedition.org/caliban/1427.

Richards, Greg and Palmer, Robert. 2010. *Eventful Cities: Cultural Management and Urban Revitalisation*. Oxford: Butterworth-Heinemann.

Robins, Kevin. 1996. *Into the Image: Culture and Politics in the Field of Vision*. London: Routledge.

Rodgers, Scott. 2020. “Journalism: An Urban Affair.” In *The Routledge Companion to Urban Media and Communication*, edited by Zlatan Krajina and Deborah Stevenson, 66-75. London and New York: Routledge.

Rose, Gillian. 2017. “Posthuman Agency In The Digitally Mediated City: Exteriorization, Individuation, Reinvention.” *Annals of the American Association of Geographers*. 107 (4): 779-793.

Ruddick, Susan. 1996. "Constructing Difference in Public Spaces: Race, Class, and Gender as Interlocking Systems." *Urban Geography* 17 (2): 132–151.

Sassen, Saskia. 2014. "Does the City Have Speech?" *City, Social Inclusion, and Education*. Barcelona: International Association of Educating Cities, 36-43.

Schiller, Naomi. 2020. "Urban Media as Infrastructure for Social Change." In *The Routledge Companion to Urban Media and Communication*, edited by Zlatan Krajina and Deborah Stevenson, 204-214. London and New York: Routledge.

Seamon, David. 1979. *A Geography of the Lifeworld: Movement, Rest and Encounter*. London: Routledge.

Sennett, Richard. 2003. "Reflections on the Public Realm." In *A Companion to the City*, edited by Gary Bridge and Sophie Watson, 380-387. Oxford: Blackwell.

Shelton, Taylor. 2017. "The Urban Geographical Imagination in the Age of Big Data." *Big Data & Society*. January–June: 1–14.

Stevenson, Deborah. 2003. *Cities and Urban Cultures*. Open University Press. Maidenhead.

Struppek, Mirjam. 2006. "The Social Potential of Urban Screens." *Visual Communication*. 5 (2): 173–188.

Therborn, Göran. 2014. "Cities as Media of Power." *Subversive Festival*, Zagreb.

Tierney, Thérèse F. 2013. *The Public Space Of Social Media: Connected Cultures Of The Network Society*. London: Routledge.

Tosoni, Simone, et al. 2019. "The Mediated City Between Research Fields: An Invitation to Urban Media Studies." *International Journal of Communication* 13: 5257–5267.

Towns, Armond R. 2020. "'The Street Is The Message': Racial Violence and the White Control of Mobility." In *The Routledge Companion to Urban Media and Communication*, edited by Zlatan Krajina and Deborah Stevenson, 437-445. London and New York: Routledge.

Vallet, Georges. 1985. "Ville Et Cite: Réflexions Sur Les Premières Fondations Qrecques En Occident." In *L'idée De La Ville*, edited by François Guéry, 56–64. Champ Vallon: Seyssel.

Valluvan, Sivamohan. 2016. "Conviviality And Multiculture: A Post-Integration Sociology of Multi-Ethnic Interaction." *Young* 24 (3): 204–221.

Vouleenaho, Jani, et al. 2015. "Digital Urbanisms: Exploring the Spectacular, Ordinary and Contested Facets of the Media City." *Observatorio* 9: http://obs.obercom.pt/index.php/obs/issue/view/48

Willems, Wendy. 2019. "'The Politics Of Things': Digital Media, Urban Space, and the Materiality of Publics." *Media, Culture & Society* 41 (8): 1192-1209.

Williams, Raymond. 1990. *Television: Technology and Cultural Form*. London: Fontana.

About the author

Zlatan Krajina is Associate Professor of Media Studies at the Department of Media and Communication, Faculty of Political Sciences, University of Zagreb. He is the author of *Negotiating the Mediated City* (Routledge, 2014) and co-editor of *The Routledge Companion to Urban Media and Communication* (2020).

9. Visualizing Locality Now: Objects, Practices and Environments of Social Media Imagery Around Urban Change

Scott Rodgers

Abstract

This chapter explores social media platforms as increasingly vast and ordinary infrastructures for how people visually experience urban locality. Drawing on a mixed-method project (including image visualization and qualitative observation) exploring the mediation of a controversial cycling scheme in East London, UK, social media imagery is conceptualized and analysed in three overlapping stages: (1) as discrete digital objects; (2) as objects mobilized through particular user practices; and (3) as objects and practices operating and appearing within platform interface environments. Drawing inspiration from phenomenological perspectives on media and technology, the chapter shows how these objects, practices and environments of social media visuality create conditions for experiencing, as well as addressing oneself towards locality, not only spatially but also temporally.

Keywords: platform urbanism, urban photography, (post)phenomenology, interface

Introduction

In everyday life, "local" is often taken for granted. It is self-evidently proximate and perhaps intimate: it is simply "here" or "there". When someone "cares" about a local place – which can range from deep worry to disinterested awareness (Dreyfus 1991, 238) – they do so on the basis of being thrown into their own local world. Stiegler's (2012) related notion of

Rose, G. (ed.), *Seeing the City Digitally. Processing Urban Space and Time.*
Taylor & Francis Group, 2022
DOI 10.5117/9789463727037_CH09

"attention", however, adds an important proviso to this kind of Heideggerian formulation about care. It suggests that a capacity to attend to any object (such as a locality) is always already exteriorized into various technical objects and infrastructures. A locality such as an urban neighbourhood is experienced as natural and effortless precisely because it is made into a durable and shared entity by countless discourses, technologies and institutions (cf. Appadurai 1995). Increasingly, these socio-technical epistemological apparatuses include digitally-mediated exteriorizations of local experience, such as gathering information on named places through search engines (Ballatore, Graham, and Sen 2017), being segregated into polygon-delineated Nextdoor neighbourhoods (Payne 2017), or knowing a nearby high street thanks partly to the efforts of Twitter-savvy local businesses (Bingham-Hall and Law 2015). When objects, activities and environments are for the most part seen as unproblematically local, what is forgotten or structurally ignored is how locally experienced actions are shaped by other local experiences and actions elsewhere, distributed across time and space.

In this chapter, I explore how the circulation of visual information through social media platforms is becoming an increasingly vast and ordinary infrastructure for how people experience urban locality. Beyond established platforms dedicated to organizing and sharing images, such as Flickr, Instagram or Pinterest, social media more generally have increasingly become important mediums of image circulation. Drawing on a collaborative research project that examined the mediation of a controversial cycling infrastructure scheme in East London, UK, I will explore the circulation of images through both Facebook and Twitter as ways of visualizing urban locality. My focus is less on the content of such images, but rather on how social media were used for producing, displaying, mobilizing and modifying various kinds of images in exchanges related to this urban controversy. For example, how through smartphone photography, users documented the installation, use, or perceived misuse of cycling infrastructures at particular locations. Or how others used editing affordances to, for example, take screenshots from online mapping services, modifying them to show planned road closures. Or how others still re-shared image-based memes that they sensed might be funny to imagined local publics.

My analysis of such social media imagery of urban change will comprise three overlapping stages. I will first conceptualize images as discrete digital objects. Second, I will examine how such objects were mobilized through particular user practices. Finally, I will consider how such objects and practices operate and appear within the interface environments of social media platforms. This threefold analysis of objects, practices and environments of social media imagery draws initially on the work of Yuk Hui, who conceptualizes digital objects as a "unity of relations" (2016, 14). Hui's relational conception of objects diverges in important respects from the object-oriented ontologies of thinkers such as Graham Harman (2002), but still provides a lens for thinking of image objects as having some autonomy from their practical use and environmental conditions. This initial focus on relatively autonomous digital image objects is important not only conceptually but methodologically, since the starting point of my analysis was visualizations of image datasets related to the above-mentioned controversy, made using ImagePlot. I will connect the patterns and properties I observe in these visualizations to qualitative observations and interviews, which revealed how such image objects were accessed, created, modified and shared through related social media practices and interface environments. Drawing inspiration from phenomenological perspectives on media and technology, I will show how these objects, practices, and environments of social media visuality create conditions for experiencing, as well as addressing oneself towards locality, not only spatially but also temporally. Social media organize and deliver visual information via temporal streams (Hochman 2014), encouraging users to engage with asynchronous image and other data alongside others, facilitating an experience of "liveness" (Van Es 2017) or "realtimeness" (Weltevrede, Helmond, and Gerlitz 2014). As a shorthand, we might say social media affords ways of visualizing locality "now".

Social platforms, imaging and urban visual experience

One way to understand social media and the city is to first situate our discussion within recent debates around the relationships of digital platforms and the city, increasingly understood by scholars as an emerging "platform urbanism" (Barns 2020; Rodgers and Moore 2018; Sadowski 2020). The meaning of "platform" is not entirely settled, in these debates or in general. As Gillespie (2010) suggests, the discursive ambiguity of the term helps platform companies take relatively equivocal stances around their own responsibilities for what happens on and through their services. Two often-interconnected meanings of platform predominate in academic debates: first, as a business model (and even a new era of capitalism) founded on extracting economic value from user contributions and metadata (e.g. Srnicek 2017); and second, as a kind of software architecture that processes and regulates data flows across external third-party websites and apps (e.g.

Helmond 2015). In the context of urban research, platforms are usually seen as relatively novel applications, services and companies that intervene and remediate a wide range of existing urban infrastructures, such as transport (e.g. Pollio 2019), food logistics (e.g. Richardson 2020), real estate (e.g. Shaw 2018), and accommodation (e.g. Wachsmuth and Weisler 2018). Arguably, there has been less attention to platforms as experiential infrastructures increasingly interdependent with how people perceive and go about daily urban life (see Barns 2019; Leszczynski 2019; Rodgers and Moore 2020; Rose et al. 2020). The approach I take here on social media visuality will emphasize in particular this experiential sense of platform urbanism.

Social media can be seen as one of the more longstanding kinds of platforms that help to produce experiences of urban locality. As their name suggests, social media cultivate social relationships and communication. Like virtually all platforms, social media individuate users by requiring them to establish a profile, using a real or invented name. What distinguishes social media from most other platforms is their emphasis on the generation, modification or circulation of content data by users. This content is valuable for social platforms since it helps in turn generate interactional metadata, relating for example to how users follow or connect with one other, share or like content, check into locations, or establish and moderate forums, groups or lists. While such content and metadata are amassed and mobilized at very large scale by social platforms, in important ways they are fundamentally local. Social data is, increasingly, location-based or spatial data (cf. Evans 2015; Gordon and de Souza e Silva 2011; Wilken 2019). In significant part this is because social media are now often accessed on the go, using smartphones and other devices. In the process, social media users reveal geospatial data concerning their usage locations, whether through automatic encoding or by volunteering it willingly. But social data are not only local in these technical or structural ways. As Loukissas argues, all data collections – including those stored on and circulated through social media platforms – should be understood as "cultural artifacts created by people, and their dutiful machines, at a time, in a place, and with the instruments at hand for audiences that are conditioned to receive them" (2019, 1–2).

Images are an increasingly important and voluminous form of social data helping to produce urban locality. The proliferation of images on social media is closely related to the rise in accessible smartphone technologies and applications, through which photography and image editing have become intimate, yet also ephemeral, affordances (Frith 2015, 88; Hand 2020). As is well known, the combination of camera phones and social platforms helps to circulate (often graphic) user-generated photographs as news imagery, which sometimes even displaces the outputs of professional photojournalism (Jukes 2018). This same combination has also enabled more ordinary forms of photographic publicity. Social media users regularly deploy photos to document, edit, and tag themselves (e.g. in selfies), banal objects, and various local wayfaring situations (Hjorth and Pink 2014; Halegoua 2019, 164–71; Hochman 2017). The sheer volume of such images, like so-called Big Data in general, should not be mistaken for the emergence of a more complete, precise or authoritative image of the city (Shelton 2017). Imaging through social media demands many of the same questions around representation that have been examined for other longstanding forms of visual media. Boy and Uitermark (2017), for example, study a dataset of more than 400,000 Instagram images related to Amsterdam, and perhaps contrary to the more ordinary or banal photography described above, find a predominance of exclusive or avant-garde places and events in the city. It is important to recognize, however, that the implications of social media for urban life extends beyond user-generated photographs. Social media users also experience, articulate, and problematize localities using re-shared and modified memes, a plethora of screen grabs, stock photography (Aiello, this volume), promotional visualizations (Rose and Willis 2019) and, increasingly, "fake" images (Highfield 2016, 92–95). The visuality fostered through social media also extends beyond discrete images: as I will argue, platform interfaces – which bring together users, images and other content – are also a way in which urban locality is visualized.

To expand on this broader perspective on social platforms, visuality, and urban experience, I will step back and very briefly consider how images emerge and are sustained as discrete digital objects through particular practices and technical environments. In *On the Existence of Digital Objects*, Hui (2016) argues that digital objects are a new kind of industrial object that has important differences with the technical objects described by Simondon (2016 [1969]), on whom she draws, alongside Heidegger (1962). Digital objects emerge through what Hui (2016: 50) calls a "double movement" of objects and data. First, data – understood in the word's original sense as things given in the world – are objectified practically. People's habitual photographic practices (Hand and Scarlett 2019), for example, or the wide range of ways they use images to establish their visibility on a social platform (Margetts et al. 2017, 137–52), are acts that formalize objects in the world by translating them into digital formats. Second, objects are datafied, and here data is understood by its contemporary meaning as computational information. This refers to the adding of attributes to an object, such as a digitally formatted image, so that it can be incorporated into a digital milieu, such as a smartphone photo browser or social media service.[1] Where digital objects are akin to the technical objects described by Simondon is that, while they can self-regulate and therefore exceed the intentions of their creators, they cannot in themselves establish new grounds of self-regulation (Ash 2018, 32). Like technical objects, digital objects are sustained as discrete entities by their "associated milieu": their physical, technical and practical conditions.

Digital images circulating through social media platforms, then, attain their discrete qualities both through users' cultural norms, practices and affective impulses, and also through how image data are continuously processed, pushed and filtered according to the technical parameters of the platform, which itself relies on data about those same users and their interactions with the platform interface. As Hochman (2014) argues, social media embody a fairly novel approach to the handling of such image objects. Rather than making images available via specific database queries, what is most distinctive about social media is the delivery of images and other data within a time-encoded data stream. This means that, experientially, social media images:

> [A]ppear to us from the current time backward and are restricted to the recent now, as older objects quickly disappear from the stream and are available only by searching the application database. What we have here is a continuous, rapid presentation of multiple data units from many users, places, and times — all appear to us almost at the same, synchronous, time. (Hochman 2014, 2)

By collating multiple images and other data from various times and places, social media create an experiential or affective "present" or "now" for variably-defined subsets of people using social media applications at particular moments (Coleman 2018). So social media images are not only interesting individually, for example in showing particular aspects of the city, or collectively, in their unprecedented volume. They are also interesting because of their environmental conditions of circulation, the streaming temporal structures of which afford new ways of experiencing urban locality.

It is important however to be cautious in describing the experience of social platforms, in which multiple visualizations of the city appear

1 Drawing on Bergson, MacKenzie and Munster (2019, 5) argue that through such datafication processes images are increasingly being rendered into large-scale ensembles of "image matter" which platforms, as socio-techncial systems, can "see" in ways distinct from human perception.

to users, as happening in "real-time". The notion of a "real-time city" has become a common, often celebratory, refrain in both popular and academic discussions around the promise of so-called smart cities (Kitchin 2014). Urban dashboards – those emerging screen interfaces which borrow their aesthetics from cars and automobiles, setting out for its users an array of urban vital signs (Mattern 2015) – are one of the prime ways in which smart cities are imagined as operating in real-time. Demonstration projects such as the London City Dashboard seem to almost scream "now!", inviting the perceiving user not only to observe but to believe in the actually existing city, delivered minute-by-minute, thanks to the data processing at work through the interface. The real-time smart city has proven to be an enchanting idea, but it has also generated critical reactions. Bleecker and Nova (2009), for example, argue that as a design principle, the efficient, seamless nature of the real-time city poses many of the same threats that the Situationists saw in modern urban planning. They propose that this might be disrupted artistically, with efforts at inserting a-synchronicity – or out-of-syncness – into urban computing.

We should go further here, however, and recognize that most flows of digital information are fundamentally asynchronous. As Sheller suggests evocatively, information flow might be seen as "less like water running smoothly down a channel, and more like an entire terrain forming in the manner of lava spreading evenly, bubbling up and overflowing, melting some structures even as it hardens into other structural forms" (2015, 19). What is important, I will suggest, is to understand that social media translate asynchronous image and other data into an apparently-real-time or synchronous experience. As Weltvrede, Helmond and Gerlitz (2014, 143) argue, events on social media do not in fact happen in real-time, in the way a synchronous videoconference does, for example. As I will discuss later, phenomena such as social media threads are obviously asynchronous; it is relatively easy to show a time series in which a photo is uploaded, then attached to a post, and then at various subsequent points commented on, reacted to and shared. What the technical operations and user practices inherent to social media forge however are specific experiential modes of "realtimeness". These temporal modes are almost infinitely multiple, because social platforms have invested in so many different kinds of tools, settings, functionalities, nudges, and features devoted to the production of a real-time-like experience (Weltevrede, Helmond, and Gerlitz 2014, 145). This real-time-like experience even paradoxically encompasses social media practices geared to memory and remembrance, such as local Facebook groups devoted to remembering places and events through curating the contribution and collection of historical photos (Keightley and Schlesinger 2014). Despite the fact that social platforms such as Facebook automatically archive image and other data, the dominant "temporal experience is one of immediacy, ephemerality, 'liveness', and flow" (Kaun and Stiernstedt 2014, 116). Stream architectures on social media continuously present new content while sending everything else downstream, immersing users in "an atmosphere and an interface of rapid change and forgetfulness" (Kaun and Stiernstedt 2014, 116).

To further flesh out this conceptualization of social platforms, imaging, and urban experience, I will now turn to consider how social media images can form part of experiencing and problematizing urban change, focusing on the mediation of a controversial cycling infrastructure scheme in Walthamstow, East London.

Social media visuality and urban change in Walthamstow, East London

In 2014, the London Borough of Waltham Forest bid for and won £30 million in funding from then Mayor Boris Johnson's "Mini Holland" programme. Run through Transport for London (TfL), this funding scheme asked for ambitious proposals to upgrade London's local cycling infrastructures. Waltham Forest developed its successful application in close collaboration with local cycling campaigners. Yet on receiving the TfL funding, the Council decided to hand a trial implementation of its scheme to its own transport engineers who, by most accounts, handled the public consultation technocratically and even clumsily. Local cycling campaigners and other supporters of the scheme soon found themselves embroiled in highly antagonistic exchanges with a range of opposing voices on Twitter and local Facebook groups. In time, these exchanges on social media became an emotional and organizational catalyst for two organized protests – one at a Town Hall vote, and a second at the scheme's official opening – that surprised the Council and attracted national media attention. As the project progressed into later phases, Waltham Forest hired a public engagement team, who led better resourced workshops, and employed Commonplace, a digital consultation platform. The scheme was also rebranded "Enjoy Waltham Forest" via a dedicated website, in an effort to pivot from cycling to a broader quality of life focus. Nevertheless, the spectre of "Mini Holland" endured and even grew as a topic of debate, promotion, ridicule, passing humour, and explanation on social media. For many it came to embody one of a series of putative divides, such as between the more and less able, young and old, or factual and emotive argumentation. For others it was an icon of encroaching gentrification, pitting longstanding, often working class and ethnically diverse residents against white middle class newcomers.

I was involved in a collaborative research project[2] focusing on three digital platforms through which publics convened around this controversy of urban change: Twitter, Facebook, and the Commonplace consultation platform. Our research design was mixed method, in that it combined and moved between qualitative and quantitative approaches to social media. Our larger-scale data analytics focused on a historical sample of just over 31,000 Twitter posts (tweets) and 11,630 Facebook posts (status updates or comments from four Public Groups and three Communities). We explored these data sets through social network analysis, topic modelling, sentiment analysis, and image visualization using ImagePlot, discussed below. These data analytics techniques were closely and recursively interwoven with qualitative techniques which included extensive observation, coding (using NVivo), and analysis of purposefully sampled online contributions and images from Facebook (including observations of five additional Private Groups), Twitter, and Commonplace, alongside twelve in-depth interviews with politicians, campaigners, activists, and frequent social media contributors. This mixed-method approach was essential since our principal interest was less in social media content than how people use social media to debate and discuss urban change, and how those uses are shaped by the technical affordances of such platforms.

The approach I take here to social media imagery is inspired by the same methodology that guided this collaborative research project. It also draws on part of its data corpus, specifically datasets of 4733 images derived from Twitter posts, and 1193 images derived from Facebook posts and comments.[3] I created a series of visualizations using the open-access application ImagePlot. These visualizations were a basis for observing patterns across

2 The research project "Planning, Participation and Social Media Platforms" was funded by an EPSRC pilot grant. The Principal Investigator was Susan Moore at University College London, and alongside myself the other co-investigator was Andrea Ballatore at Birkbeck, University of London, who I must also thank for helping prepare the image data sets used in this chapter.

3 Most of these images were downloaded via a sample dataset of social media posts, scraped through the Twitter and Facebook APIs by Andrea Ballatore. The original social media posts were drawn from four time periods between 2014 and 2017. Our doctoral research assistant Justinien Tribillon conducted a preliminary analysis of Facebook and Twitter, as well as news coverage, to identify the time period with the most intense activity. I added to the scraped image dataset by manually downloading 580 images from closed Facebook groups.

large collections of images, and then identifying areas for closer investigation (Hochman and Manovich 2013). Unlike some digital humanities approaches, this closer analysis was not focused on analysing the meaning of individual images. Instead, I moved between the image visualizations and selected qualitative observations of how such images figured into practical and environmental contexts such as posts or threads. I also cross-referenced these observations with themes coded in interview transcripts. Thus, across the next three subsections, the volume of images I bring into view will go from larger to smaller. However, my analytical scope will also widen, from images as discrete objects, to the practices through which they circulate, and then, to some of the dynamics of their interface environments.

Objects

I will begin by treating the images I visualized somewhat narrowly, as sets of discrete digital objects. First, I need to provide just a little more detail about how I used ImagePlot to create these image visualizations.[4] ImagePlot is an open-access software application created by the Software Studies Initiative, which runs as a macro within ImageJ. It renders arrays of images onto a two-dimensional canvas, delimited by an XY axis, along either cartesian or polar coordinates. The image values ImagePlot uses are set by the user, and in my case, I captured and used two main types of data. First, data related to the dimensions and colouration of the images (median and standard deviation saturation, hue and brightness), measured using standard ImagePlot macros. Second, image metadata collected through the Facebook and Twitter APIs as part of the collaborative project mentioned earlier. This metadata related to: the date, weekday, and time an image was posted; the number of "reactions" (e.g. likes, retweets, comments); and in the case of the Twitter image data, the followers and friends count of the tweet author.[5] We might observe that the first type of image data more closely reflects Hui's (2016, 50) first "movement", where objects in the world are translated into digital formats, whereas the second are more related to data that codes objects (here, images) into a digital milieu (i.e. a social media platform).

4 Readers seeking more detail about the functions and various uses of ImagePlot should visit the Software Studies Initiative website, at which there is extensive documentation and related literature. See: http://lab.softwarestudies.com/p/imageplot.html

5 Due to restrictions in the Facebook API, we were only able to collect such metadata for posts made to public groups. Additionally, while the Twitter API allowed us to collect data on the retweet count for each post, it did not at the time of collection provide the number of "likes".

While I did not create visualizations for every possible combination of the above data points – there were hundreds of possible combinations – I took a relatively open-ended and experimental approach, producing eighty-nine visualizations, split evenly between the Twitter and Facebook datasets, as well as between plots along Cartesian or polar coordinates. I created such a large number of visualizations for a few reasons. Experimenting with different visualizations was necessary to learn how to use ImagePlot effectively, particularly since I am primarily a qualitative researcher, lacking experience in data analytics techniques. Consequently, I was also probably motivated to justify the considerable effort I had put into preparing and cleaning the required data tables. However, in the process of creating the visualizations I also began to understand that some of the most interesting patterns were emerging more by accident than design, which encouraged me to try different combinations of data variables.

Contrary to my expectations, visualizations plotting the image metadata, for example, the number of image reactions (e.g. "likes", angry faces, retweets), were not very revealing. In the case of reactions, there was too little variation. Only a small number of posts garnered voluminous reactions; most had only a few, and many had none at all. On closer investigation, larger numbers of reactions appeared to be influenced by a range of factors, including but beyond the image used. For example, one image on Facebook with 139 comments was attached to a post offering a heartfelt testimony about the negative impact the Mini Holland scheme on the author's long-time business on Orford Road in Walthamstow. The image, showing the part-pedestrianization of Orford Road, empty of cars but otherwise arguably banal, was only loosely related to the post. Another image on Twitter, with 386 retweets, is attached to a post proclaiming: "Astonishing results from Walthamstow Mini Holland: 10,000 fewer cars a day, traffic down 56% in village, no collisions." The image relates directly to the claim: a photographed page from a Council report, with substantiating passages underlined in pen. However, the number of reactions is also likely because the tweet was authored by Ross Lydall, Health and City Hall Editor for the *London Evening Standard*. Working for a large mainstream newspaper, he possesses status and authority, with over 8000 followers at the time of posting.

The point is not that these examples are uninteresting, particularly not when put into their practical and environmental contexts. Nor do I mean to deny the important role such metadata plays in providing a ground that makes such images specifically digital objects. However, at the scale of visualizing a large collection of such image objects, colouration and size data showed more interesting patterns. Figure 9.1, for example, shows part

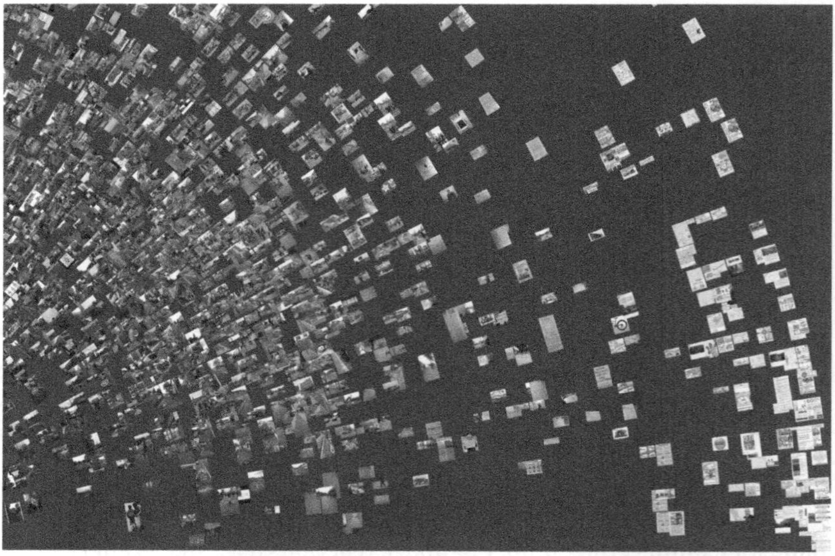

Figure 9.1. Cropped section of ImagePlot render of 4733 Twitter images, plotting brightness (X axis) against saturation (Y axis), along polar coordinates. Made by author.

of a visualization from our Twitter dataset, plotting image brightness (X) against saturation (Y). Towards its left side, the figure is dominated by daytime images that include grey pavement. These represent a prevalent "infrastructure photography" that I discuss further in the next section. The cluster of white background images towards the right side also jumps out, however. These images are predominantly screengrabs from sources such as social media posts, campaigning material, news websites and reports. Material like this was not surprizing; our social network analysis showed that cycling campaigners and their allies (e.g. local councillors, academics) dominated Twitter discussions around Mini Holland, and might be expected to share material like this. The volume of these kinds of images was however more unexpected, and served as a basic but important reminder that social media images in our dataset – and probably more generally – very often include a wide variety of materials beyond photographs. Visualizations of colouration patterns plotted the properties of images as such, rather than the attributes allowing their circulation within a social platform. In so doing, these revealed how exchanges about the Mini Holland programme produced an "extensive, diverse and febrile visual field" of image objects, experienced as often through "glancing at a screen, swiping quickly through networked images, briefly scanning a feed" (Rose and Willis 2019, 422) as focused attention.

Visualizations plotting image dimensions were also revealing, showing some of the different modalities through which users in our case studies captured, sometimes edited and shared images. There are clusters of images at common photo proportions (2:3/3:2, 4:3/3:4, as well as 1:1, often associated with Instagram). On closer inspection, a prevalence of smartphone photography amongst these images is clear (as discussed in the next section), even though this could not be plotted, since both the Facebook and Twitter APIs strip most of the image EXIF data, which can indicate the camera used. Another large cluster of images can be seen at 9:16/16:9, the most common aspect ratio for smartphone screens. In most cases, these images were likely created from a smartphone screengrab, for example capturing and re-sharing an image of a social media post, part of a comment thread, or part of a news story. Beyond these typical image proportions are a range of slender (wide and thin) images, which includes banners, logos, calls to action, and partial screengrabs of quoted text (e.g. from a report, or social media). In some cases, these images might have been ready-cropped and re-shared. In others, the user likely cropped the image themselves, indicating some digital literacy, and possibly editing on a desktop computer. In the Twitter image collection, in particular, there is evidence of more variation in image proportions, possibly suggesting users more predisposed, comported or in a situation (such as a desk-based job) to precisely crop images they share.

Digital images are "objects" in that they anchor and sustain the technical operations of social platforms, as well as how users navigate platform interfaces in the contexts of their wider everyday existence. Even considered in relative isolation, the visualizations I created show a field of objects helping to generate the "phenomenological surface of sensemaking" (Pentzold and Menke 2020, 2804) increasingly presented through social media environments. So far, however, these ImagePlot visualizations have comprised a more present-at-hand orientation to the collected images (to use Heideggerian terminology, cf. Hui 2016, 16): an analysis in other words trained on the image collections as such, visually rendered based on their measurements and social metadata. Let me turn now to extend my analysis into these images as ready-to-hand: that is, as objects receding into the background of different practices in and through social media.

Practices

It may already be evident that divorcing image objects from their practical contexts is difficult. In our context, "practices" can refer to very fine-grained, dispersed understandings or rules (or "doings" and "sayings", see Schatzki [2002]) around using, noticing, capturing, editing, and sharing images in and through social media. Here, however, I am also interested in how images figure into more complex forms of practical purposiveness or end orientation, ranging from the more teleological or normative to the more affective or emotional (Schatzki 2002, 80–83). One more obvious form of organized image-related practices found in exchanges around the Mini Holland controversy was campaigning. Earlier Twitter posts, for example, include many images related to calling others to action, such as banners, posters and digital flyers. These images tend to appear in posts that relate to broader campaigning around cycling. Let me focus here, however, on the wide range of imaging practices we observed that oriented to the controversy as a specifically local concern.

One clear manifestation of visualizing urban locality were various practices of witnessing the Mini Holland programme as it was implemented in physical spaces, similar to the acts of mobile wayfaring described by Hjorth and Pink (2014). Although, as discussed in the previous section, detailed EXIF data is unavailable in our image dataset, smartphones (or camera phones) can reasonably be inferred as the prevailing technology to hand in such witnessed photographs. Not only because of the image proportions, but evident photographic perspectives and situations, such as from inside a car stuck in traffic, or shot while walking or cycling, or of documents laying upon a table, just received in the post. Photos like these imply a camera angle or position, and an act of photography, that would have only been possible, or at least be more likely, with the slender size and social discreetness of a smartphone (Bate 2013, 81).

The most common "witnessing" smartphone photographs capture under-construction or recently finished cycling infrastructures at the centre of the controversy – or otherwise their absence. This "infrastructure photography" was not only evident in our qualitative observations, but also in the ImagePlot visualizations. Plots of colouration show large grey-toned clusters of images with the asphalt or stone paving slabs involved in road and cycling infrastructure, often newly installed and purer in colour. Also visible are image clusters of reddish-orange with restricted entry signs or construction barriers, yellow with road construction notices, red bike lanes, green bicycle storage sheds, and dark anthracite bollards. This kind of witnessed infrastructure photography was put to a variety of purposes. One of the most active contributors in our dataset, on both Facebook and Twitter, and a supporter of the cycling scheme, described his use of mobile photography to us in this way:

> I use photos almost like a note. So if I see something I want to discuss with anyone about what's happening, something that's going on, something I think should change or I think it's bad, well to do with anything really, I will take a photograph of it and then when I'm going through my photos, yes then I might do a post to say, 'I saw this the other day'.

Activists and other proponents often used photography of new cycling infrastructure, or related environmental improvements such as commissioned street art, to call attention to and even celebrate the Council's scheme. One of the most photographed locations in that vein was Orford Road, perhaps Walthamstow's most gentrified street, and the site of the Mini Holland project's first phase and official opening. A leading member of the Waltham Forest Cycling Campaign described his own photography of the street:

> Well, I guess quite a lot of my tweeting is around, you know, what I see in the street, which I think looks good, looks nice, looks comfortable. So in a way I'm just as guilty of that in a way, you know, my shots of Orford Road of a particularly nice crossing or whatever it might be that I'm seeing is like, that's a feel-good thing about, yes, you know, stuff is starting to look better, you know. Look at the plants or whatever it might be [...] it's emotional language, isn't it?

This "emotional language" was also visible in visualizations plotting colouration. In infrastructure shots with strong blue tones, the clear sky that often occupies half the frame was possibly an aesthetic cue for the photographer. High saturation green images, meanwhile, show a variety of summer events in public parks, often with happy cycling children riding on the grass.

Celebration or aesthetic appreciation were not the only ways images embodied an emotional language. For many, particularly on Facebook, images and their ensuing discussion threads centred on venting: the expression of how one feels, and in so doing the releasing of pressure. Venting related to the Mini Holland project could be about a range of things, such as bad drivers, inconsiderate cyclists, the congested school run, unused cycle lanes, or damaged infrastructure. Most commonly, venting related to traffic delays, evident for example in visualizations plotting brightness against hue, where clusters of images show evening traffic scenes, such as a line of glaring headlights cutting through dusk or dark. Users posting these images frequently deployed the term "road chaos" and directly ascribed such traffic to the Mini Holland programme, with the image itself serving as evidence.

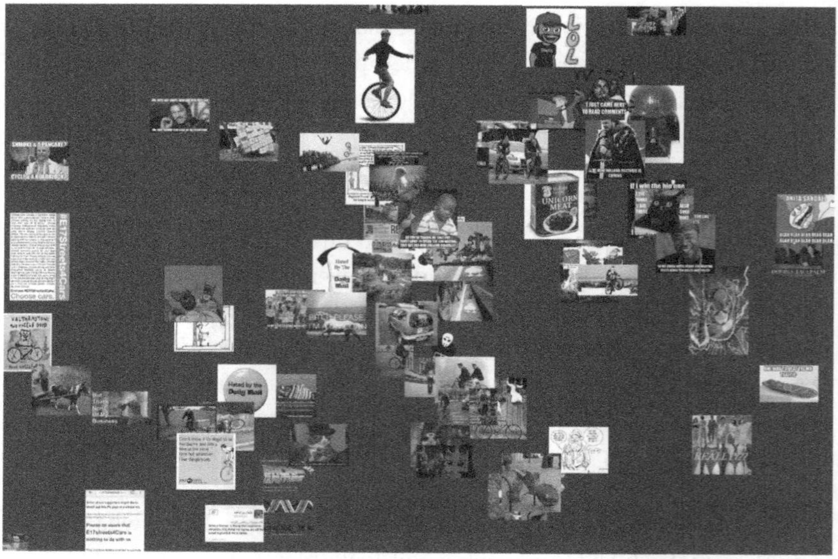

Figure 9.2. Cropped section of ImagePlot render of 85 Facebook and Twitter images (purposefully sampled examples of humouristic use), plotting standard deviation hue (X axis) against standard deviation saturation (Y axis), along cartesian coordinates. Made by author.

The recurring invocation of "road chaos" also catalysed humoristic riffing and banter, often augmented by images, which subtly demarcated users according to whether they were in on the joke or not. For example, there was a minor memetic subgenre in the Facebook groups we studied of users parodying others that blamed Mini Holland for increased traffic, offering absurd accounts of the scheme's effects on other happenings, from the weather, to mobile phone mast construction, to the lateness of Ocado food deliveries. In addition to these kinds of memetic jokes, users also used images as such to convey a point humorously (shown in Figure 9.2), including various memes such as image macros (a recurring image with a superimposed caption).

In contrast to exasperated venting, or banter with like-minded others, Facebook and Twitter users also often deployed images to illustrate or substantiate claims. Many of these are the same white background screengrabs I mentioned in the last section – slices of social media posts, campaigning material, news websites or reports (e.g. of passages, photographs, charts, tables, maps) – used to illustrate a point, or cite an authoritative source. Sometimes these screengrabs included additional editing, for example maps edited to delimit areas, highlight routes, show distances, or pinpoint locations with various qualities. Photographs were also sometimes used as a kind of evidence. For instance, some Facebook users photographed printed material such as correspondence or public notices received in the post from Enjoy Waltham Forest, or less frequently, newspaper pages to mark a new development.

Of course, the infrastructure photography I mentioned earlier was the most common forms of visual evidence or substantiation. Here, there were some notable differences between such infrastructure photography on Facebook and Twitter. In Facebook neighbourhood groups, users tended to present themselves within the context of their everyday life, while showing an awareness that they share local turf with their interlocutors. Their use of photography relating to the Mini Holland scheme tended to accordingly be highly local examples of infrastructural objects or situations, often characterized by shots of bare infrastructure without people, sometimes closeup. On Twitter, users tend to adopt more of a public persona, and share content to broadcast a perspective, form networks and partake in broader and often translocal conversation with others (cf. Binns 2014). Infrastructural images on Twitter often exemplified good design or policy "best practices", and were characterized by wide shots showing people using such infrastructure (e.g. walking, cycling, or sitting on a pedestrianized pavement). Sometimes, these images were skilfully shot images of Mini Holland infrastructure, but there were also more generic photographs, often re-shared repeatedly, showing international examples of intersection design, bike lanes, or public realm improvements, as well as design visualizations and staged photos (e.g. comparisons of road usage by pedestrians, cyclists, buses and cars). These are not just differences in user practices, but also in the characteristics of the platforms' different interface environments, which I now turn to discuss.

Environments

Just as it is difficult to separate image objects from their practical instantiation, it is also difficult to separate imaging practices from their socio-technical environments. By "environment" I primarily mean the digital interfaces of social media platforms. Digital interfaces can be approached in a number of ways. Ash (2015, 16–24), for example, summarizes five distinct if overlapping approaches to digital interfaces, as computational, practiced, ideological, an analogue/digital bridge, and affective. There is no space here to delve into these various conceptual debates around interfaces, nor Ash's own (2015, 25–32) object-oriented approach. Suffice to say, I will approach social media interfaces here as an irreducible duality. On the one hand, they are experiential infrastructures, defined not so much by what they are, but rather what they do in conjunction with users' practices (Rodgers and Moore 2020). On the other hand, they are technical infrastructures which govern the relationships and presentation of various digital objects, such as user profiles, image and other content, and interactional affordances such as threads.

Rather than considering social media interfaces abstractly, I will examine one aspect that was important in our study: threaded exchanges, which are core to how social media organize and deliver images alongside other content in temporal streams. Let us start with a closer if brief account of just one threaded exchange. It is drawn from the Walthamstow Life Facebook group, the second largest carrying the Walthamstow name. The thread begins with an image. Likely taken with a smartphone, it shows Orford Road, as mentioned earlier the symbolic epicentre of the Mini-Holland controversy. People are seated as well as standing outside some of the road's cafes on a partly sunny day. They are somewhat hemmed in by temporary orange construction barriers. Compositionally, it is not a work of art, nor would that be expected. But Justin,[6] the user that took the photo, seems to know the photographic "rule of thirds" and how to work the basics of perspective. The newly expanded sidewalk fills the bottom of the frame, vanishing into the upper-left third of the image.

Justin's image includes a textual comment: "Good to see local people and businesses on Orford Road flourishing from the Mini Holland enhancements." A few minutes later, he adds another comment to his relatively neutral first one, acknowledging the different views about the cycling scheme, but suggesting everyone can benefit from improved public space. He exhorts people to get more involved and make the most of the infrastructure. At first, there are just a few banal comments about the image, such as the pavement width and the construction barriers. Soon, however, a familiar set of antagonistic exchanges build up around the Mini Holland programme: that not all shops benefit from pedestrianization; that more trees are needed; that the council just cannot be trusted. Flo, one of the users who had initially raised some of the earlier critical questions, returns to the discussion to report that she's seen on another local Facebook group that a cyclist has been run over by a bus. "Maybe the mini Holland money should have been spent on busy roads," says Flo, "where we have accidents,

6 Users named in this account have been given pseudonyms. In addition, quoted passages from our dataset have been reworded, while remaining faithful to the original expression, in order to further cloak the users and protect their anonymity.

not orford rd, one of E17's safest areas." "Please tell me this isn't going to be one of *those* threads," bemoans Kamron, a well-known cycling campaigner, who beckons people to look at all the detailed information on the Enjoy Waltham Forest website about the full range of infrastructure improvements planned. Another, Rahila, accuses Flo of cynically using a potentially injured cyclist to prove a point.

Just as the exchange heats up, it is momentarily interrupted by another that seems to take everything back to the original image:

> Charlotte: I saw u snap that pic earlier
> Justin: Sorry, Charlotte, are you in it? Should I remove it? I tried to shoot the photo without faces....
> Charlotte: Naw dnt b daft it's my corner shop, the bakery!
> Justin: Oh wait I know u!
> Justin: Whoops, hi Lotta!
> Charlotte: Hey!! How'd you like the sandwich?
> Justin: Tikka Masala Chicken was great!

After this amusingly hyperlocal interlude, a long and often heated back and forth discussion gains momentum again around the Mini Holland programme. Beyond discussions of the cycling infrastructure, personal attacks are hurled between different users, some of whom are already known to one another. Stipulations and arguments are also made around etiquette on the thread itself, such as whether cutting and pasting past comments is appropriate, or whether a pointed discussion should just continue through "pm" (personal messaging).

One way to see the scale of a thread like this, typical in our case study, is to zoom out and visualize its entirety, as a series of screengrabs (Figure 9.3). In the mode of clock time, measured from initial post to last comment, Justin's post and image inaugurated a thread of 142 comments, spread across two days (forty-nine hours, or just one shy of 3000 minutes). The thread is striking not just in scale or duration, but as a visual representation of the asynchronicity of threaded exchanges, when mapped according to their occurrence in clock time. But what is most interesting on closer analysis is how users' contributions to such asynchronous, threaded exchanges express a real-time-like (Weltevrede, Helmond, and Gerlitz 2014) and largely transitory (Kaun and Stiernstedt 2014) experience. The posting of the image, observations of its details, critical interjections about the cycling scheme's equity, importing news from adjacent Facebook groups, interludes about bakery sandwiches, debates on thread etiquette: these embody not a single "now" but a succession of "nows".

Figure 9.3. Visualization of threaded Facebook exchange (1 post with image, 142 comments) captured using multiple screengrabs. Made by author.

This threaded succession of nows illustrates the irreducible duality of platform environments, as both experiential infrastructures produced through user practices, and technical infrastructures governing user relationships with each other, as well as with other digital objects. The commenting practices, as well as Justin's initial image, are significant, but the thread is not just the sum total of these asynchronous acts. As an encompassing visual environment, the thread itself encourages contributions. Social media interfaces, and threads in particular, are designed to generate anticipation of further updates and responses. They can be understood as a kind of animation, not just in the general sense of bringing things to life, but more specifically in creating a visual appearance of movement. Both Facebook and Twitter animate contributions through time-based threads, though with differing forms of publicity and intimacy. The Facebook interface encourages users to delve into relatively focused threaded discussions, and has been deepened recently, with added functionalities such as nested (hierarchical) threads and a wider series of emotional reactions (e.g. thumbs up, angry face, laughter, etc). The Twitter interface, meanwhile, prioritizes reacting and contributing to the main user feed, for example by presenting replies within the same visualized time stream as initial tweets. In the context of our study, which focused on social media contributions made broadly in relation to a shared geographical context, the platform interface is therefore more than just a delivery system for discrete images of or related to an urban locality. With users engaging social platforms through all manner of places or settings (often using mobile devices), different streaming interfaces should be understood as visual environments in their own right – as part of experiencing and seeing the "near" and "now" of urban locality.

Conclusion

In some ways my analysis of social media visuality and the city in this chapter has been unconventional, combining phenomenological perspectives on media and technology, qualitative research, and computational approaches to image visualization. I began with an exploration of digital image collections, visualized using ImagePlot, because it was methodologically as well as conceptually useful. The visualization of image objects showed both discrete crystallizations of users' cultural norms, practices and affective impulses, and at the same time, discrete data units that are processed, pushed and filtered through technical platform environments. In both respects, I was able to observe the images as generative objects.

Digital images are inseparable from the practices and performances that go into, for example, making or taking a picture, or the technical conditions that allow them to circulate across different platforms. Beginning with these image objects could be seen, following Kember and Zylinska (2012, 71–95), as homing in on a specific "cut" in the flow of mediation, one that is technical and material but also perceptual and affective. While I had some interest in how these images documented past moments, or marked the passage of time, my prime concern was their lively and productive dimensions.

One of the main ways I explored the "liveliness" of such images was their figuration within practices of social media contribution around the Walthamstow Mini Holland controversy, as a specifically local concern. Chief amongst such practices were photographic acts of witnessing the cycling programme as it unfolded through new infrastructure and related elements – celebrating and aesthetically appreciating the scheme and its implementation, or venting and trading jokes about it. But in so doing users also deployed a range of other images beyond photography, from memes to various visual materials meant to illustrate or substantiate claims. Whether these imaging practices were devoted to rational deliberation, or just exasperated venting or banter with like-minded others, they can all be seen as "ambient" forms of participation around urban transformation: participation that is as often dispersed and incidental as it is discrete or directed. This need not be seen as cluttered or obfuscated engagement in local life. At least in principle, participation through more ambient forms of social media visuality can "invite tuning in instead of tuning out" (McCullough 2013, 20).

The urban visuality of social media in this chapter has not just referred to images, but also platform interfaces. Digital interfaces are not so much a visualization of urban spaces understood representationally, but visual environments that are increasingly part of the urban experience. My account analysed social media streams, and specifically threads, as both experiential (Rodgers and Moore 2020) as well as technical infrastructures in this respect. Contributing to a streamed interface such as a thread involves partaking in an environment that is structurally asynchronous but experientially future-oriented; a kind of visual animation between the movements of the platform and users. In a context where digital content and services are increasingly accessed through mobile interfaces – which recede more and more into the background of their users' situations – social media and their images appear to bring about novel spatially- and temporally-organized environments for seeing, knowing and living in cities.

References

Appadurai, Arjun. 1995. "The Production of Locality." In *Counterworks: Managing the Diversity of Knowledge*, edited by R. Fardon, 208–29. London: Routledge.

Ash, James. 2015. *The Interface Envelope: Gaming, Technology, Power.* New York: Bloomsbury.

———. 2018. *Phase Media: Space, Time and the Politics of Smart Objects*. New York: Bloomsbury.

Ballatore, Andrea, Mark Graham, and Shilad Sen. 2017. "Digital Hegemonies: The Localness of Search Engine Results." *Annals of the American Association of Geographers* 107 (5): 1194–1215.

Barns, Sarah. 2019. "Negotiating the Platform Pivot: From Participatory Digital Ecosystems to Infrastructures of Everyday Life." *Geography Compass* 13 (9): https://doi.org/10.1111/gec3.12464.

———. 2020. *Platform Urbanism: Negotiating Platform Ecosystems in Connected Cities*. London: Palgrave Macmillan.

Bate, David. 2013. "The Digital Condition of Photography: Cameras, Computers and Display." In *The Photographic Image in Digital Culture*, edited by Martin Lister, 91–108. London: Routledge.

Bingham-Hall, John, and Stephen Law. 2015. "Connected or Informed?: Local Twitter Networking in a London Neighbourhood." *Big Data & Society* 2 (2): https://doi.org/10.1177/2053951715597457.

Binns, Amy. 2014. "Twitter City and Facebook Village: Teenage Girls' Personas and Experiences Influenced by Choice Architecture in Social Networking Sites." *Journal of Media Practice* 15 (2): 71–91.

Bleecker, Julian, and Nicolas Nova. 2009. "A Synchronicity: Design Fictions for Asynchronous Urban Computing." In *Situated Technologies Pamphlets*, edited by Omar Khan, Trebor Scholz, and Mark Shepard. Vol. 5. New York: The Architectural League.

Boy, John D, and Justus Uitermark. 2017. "Reassembling the City through Instagram." *Transactions of the Institute of British Geographers* 42 (4): 612–24.

Coleman, Rebecca. 2018. "Social Media and the Materialisation of the Affective Present." In *Affect and Social Media: Emotion, Mediation, Anxiety and Contagion*, edited by Tony D. Sampson, Stephen Maddison, and Darren Ellis, 67–75. London: Rowman & Littlefield International.

Dreyfus, Hubert L. 1991. *Being-in-the-World: A Commentary on Heidegger's Being and Time, Division I*. Cambridge, MA: MIT Press.

Es, Karin van. 2017. "Liveness Redux: On Media and Their Claim to Be Live." *Media, Culture & Society* 39 (8): 1245–56.

Evans, Leighton. 2015. *Locative Social Media*. London: Palgrave Macmillan.

Frith, Jordan. 2015. *Smartphones as Locative Media*. Cambridge: Polity.

Gillespie, Tarleton. 2010. "The Politics of 'Platforms'." *New Media & Society* 12 (3): 347–64.

Gordon, Eric, and Adriana de Souza e Silva. 2011. *Net Locality: Why Location Matters in a Networked World*. Chichester: Wiley-Blackwell.

Halegoua, Germaine R. 2019. *The Digital City: Media and the Social Production of Place*. New York: New York University Press.

Hand, Martin. 2020. "Photography Meets Social Media: Image Making and Sharing in a Continually Networked Present." In *The Handbook of Photography Studies*, edited by Gil Pasternak, 310–326. London: Bloomsbury.

Hand, Martin, and Ashley Scarlett. 2019. "Habitual Photography: Time, Rhythm, and Temporalization in Contemporary Personal Photography." In *The Routledge Companion to Photography Theory*, edited by Mark Durden and Jane Tormey, 410–25. New York: Routledge.

Harman, Graham. 2002. *Tool-Being: Heidegger and the Metaphyics of Objects*. Chicago: Open Court.

Heidegger, Martin. 1962. *Being and Time*. Oxford: Blackwell.

Helmond, Anne. 2015. "The Platformization of the Web: Making Web Data Platform Ready." *Social Media + Society* 1 (2): https://doi.org/10.1177/2056305115603080.

Highfield, Tim. 2016. *Social Media and Everyday Politics*. Cambridge: Polity.

Hjorth, Larissa, and Sarah Pink. 2014. "New Visualities and the Digital Wayfarer: Reconceptualizing Camera Phone Photography and Locative Media." *Mobile Media & Communication* 2 (1): 40–57.

Hochman, Nadav. 2014. "The Social Media Image." *Big Data & Society* 1 (2): https://doi.org/10.1177/2053951714546645.

———. 2017. "From Site-Specificity to Hyper-Locality: Performances of Place in Social Media." In *The SAGE Handbook of Social Media Research Methods*, edited by Luke Sloan and Anabel Quan-Haase, 367–85. London: Sage.

Hochman, Nadav, and Lev Manovich. 2013. "Zooming into an Instagram City: Reading the Local through Social Media." *First Monday*, June: https://doi.org/10.5210/fm.v18i7.4711.

Hui, Yuk. 2016. *On the Existence of Digital Objects*. Minneapolis: University of Minnesota Press.

Jukes, Stephen. 2018. "The Newsroom Is No Longer a Safe Zone: Assessing the Afffective Impact of Graphic User-Generated Images on Journalists Working with Social Media." In *Affect and Social Media: Emotion, Mediation, Anxiety and Contagion*, edited by Tony D. Sampson, Stephen Maddison, and Darren Ellis, 168–77. London: Rowman & Littlefield International.

Kaun, Anne, and Fredrik Stiernstedt. 2014. "Facebook Time: Technological and Institutional Affordances for Media Memories." *New Media & Society* 16 (7): 1154–68.

Keightley, Emily, and Philip Schlesinger. 2014. "Digital Media – Social Memory: Remembering in Digitally Networked Times." *Media, Culture & Society* 36 (6): 745–47.

Kember, Sarah, and Joanna Zylinska. 2012. *Life after New Media: Mediation as a Vital Process*. Cambridge, MA: MIT Press.

Kitchin, Rob. 2014. "The Real-Time City? Big Data and Smart Urbanism." *GeoJournal* 79 (1): 1–14.

Leszczynski, Agnieszka. 2020. "Glitchy Vignettes of Platform Urbanism." *Environment and Planning D: Society and Space*, 38 (2): 189–208.

Loukissas, Yanni A. 2019. *All Data Are Local: Thinking Critically in a Data-Driven Society*. Cambridge, MA: MIT Press.

McCullough, Malcolm. 2013. *Ambient Commons: Attention in the Age of Embodied Information*. Cambridge, MA: MIT Press.

MacKenzie, Adrian, and Anna Munster. 2019. "Platform Seeing: Image Ensembles and Their Invisualities." *Theory, Culture & Society* 36 (5): 3–22.

Margetts, Helen, Peter John, Scott A. Hale, and Taha Yasseri. 2017. *Political Turbulence: How Social Media Shape Collective Action*. Princeton: Princeton University Press.

Mattern, Shannon. 2015. "Mission Control: A History of the Urban Dashboard." *Places Journal* (March): https://doi.org/10.22269/150309.

Payne, Will. 2017. "Welcome to the Polygon: Contested Digital Neighborhoods and Spatialized Segregation on Nextdoor." *Computational Culture* 6.

Pentzold, Christian, and Manuel Menke. 2020. "Conceptualizing the Doings and Sayings of Media Practices: Expressive Performance, Communicative Understanding and Epistemic Discourse." *International Journal of Communication* 14: 2789–2809.

Pollio, Andrea. 2019. "Forefronts of the Sharing Economy: Uber in Cape Town." *International Journal of Urban and Regional Research* 43 (4): 760–75.

Richardson, Lizzie. 2020. "Platforms, Markets, and Contingent Calculation: The Flexible Arrangement of the Delivered Meal." *Antipode* 52 (3): 619–36.

Rodgers, Scott, and Susan Moore. 2018. "Platform Urbanism: An Introduction." *Mediapolis: A Journal of Cities and Culture* 4 (3): Available at: http://www.mediapolisjournal.com/2018/10/platform-urbanism-an-introduction/.

———. 2020. "Platform Phenomenologies: Social Media as Experiential Infrastructures of Urban Public Life." In *Urban Platforms and the Future City: Transformations in Infrastructure, Governance, Knowledge and Everyday Life*, edited by John Stehlin, Michael Hodson, Julia Kasmire, Andrew Mcmeekin, and Kevin Ward, 209-222. London: Routledge.

Rose, Gillian, and Alistair Willis. 2019. "Seeing the Smart City on Twitter: Colour and the Affective Territories of Becoming Smart." *Environment and Planning D: Society and Space* 37 (3): 411–27.

Rose, Gillian, Parvati Raghuram, Sophie Watson, and Edward Wigley. 2020. "Platform Urbanism, Smartphone Applications and Valuing Data in a Smart City." *Transactions of the Institute of British Geographers*, July. https://doi.org/10.1111/tran.12400.

Sadowski, Jathan. 2020. "Cyberspace and Cityscapes: On the Emergence of Platform Urbanism." *Urban Geography* 41 (3): 448–52.

Schatzki, Theodore R. 2002. *The Site of the Social: A Philosophical Account of the Constitution of Social Life and Change*. University Park: Penn State University Press.

Shaw, Joe. 2018. "Platform Real Estate: Theory and Practice of New Urban Real Estate Markets." *Urban Geography* 41 (8): 1037-1064.

Sheller, Mimi. 2015. "News Now: Interface, Ambience, Flow, and the Disruptive Spatio-Temporalities of Mobile News Media." *Journalism Studies* 16 (1): 12–26.

Shelton, Taylor. 2017. "The Urban Geographical Imagination in the Age of Big Data." *Big Data & Society* 4 (1): https://doi.org/10.1177/2053951716665129.

Simondon, Gilbert. 2016. *On the Mode of Existence of Technical Objects*. Translated by Cecile Malaspina and John Rogove. Minneapolis: Univocal Publishing.

Srnicek, Nick. 2017. *Platform Capitalism*. Cambridge: Polity.

Stiegler, Bernard. 2012. "Relational Ecology and the Digital Pharmakon." *Culture Machine* 13: 1–19.

Wachsmuth, David, and Alexander Weisler. 2018. "Airbnb and the Rent Gap: Gentrification through the Sharing Economy." *Environment and Planning A: Economy and Space* 50 (6): 1147–70.

Weltevrede, Esther, Anne Helmond, and Carolin Gerlitz. 2014. "The Politics of Real-Time: A Device Perspective on Social Media Platforms and Search Engines." *Theory, Culture & Society* 31 (6): 125–50.
Wilken, Rowan. 2019. *Cultural Economies of Locative Media*. Oxford: Oxford University Press.

About the author

Scott Rodgers is Reader in Media and Geography in the Department of Film, Media and Cultural Studies at Birkbeck, University of London. His research specializes in the relationships of media and cities and the geographies of communication. Scott also has broad interests in social media, journalism, urban politics, phenomenological approaches to media and technology, and ethnographic methodologies. His most recent work focuses on the convergence between digital platforms and the making of contemporary urbanism and locality.

10. Perfect Strangers in the City: Stock Photography as Ambient Imagery

Giorgia Aiello

Abstract
In this chapter, I focus on stock photography as an unremarkable and mundane visual genre that permeates everyday life in cities, and which therefore also contributes to shaping experiences in and of urban public space. The chapter conceptualizes stock photography as an ambient medium. It shows how stock photography is used to promote businesses and services through smiling individuals and other familiar and flexible subjects that often enliven otherwise vacant or drab storefronts and ultimately also confer warmth to otherwise bland if not alienating urban "landscapes of capital". The chapter concludes by reflecting on some of the problems and potentials of this often overlooked type of ambient imagery in the creation and regulation of urban mood.

Keywords: stock photography, atmospheres, urban mood, the everyday

Introduction

Stock visuals are everywhere, though we rarely notice them – not unlike the air we breathe. They include infographics, illustrations, video footage and, perhaps more predictably, also creative and editorial photographs that can be licensed from global visual content providers like Getty Images, Shutterstock and many others. These pre-produced visuals, and especially stock photography, are the visual bread and butter for much of contemporary communication. With the rise of social networking and online news, stock photos have become central to digital media culture. Far from being only encountered in the digital realm, however, stock photography is also the raw material of much visual communication in urban space. This is not only because outdoor advertising

Rose, G. (ed.), *Seeing the City Digitally. Processing Urban Space and Time.*
Taylor & Francis Group, 2022
DOI 10.5117/9789463727037_CH10

is increasingly made of layouts that integrate stock imagery into narratives promoting branded commodities, but also and more importantly because stock photography is now regularly displayed in public space independently of traditional advertising as such. This is a topic that remains under-researched, but which could in fact shed some light on how the proliferation of digitally produced and distributed imagery shapes our everyday urban surroundings.

In this chapter, then, I focus specifically on stock photography's presence in the urban built environment, with an emphasis on some of the ways in which generic images contribute to the promotion of particular experiences in and of public space. In doing so, I take on Paul Frosh's recent invitation to move beyond the "standard critique" of stock photography as being clichéd and stereotypical, and therefore also problematic from a representational standpoint, to gain insight into its characteristics "as an ambient or 'absent-minded' medium" (2020, 189), which is just as banal as it is abundant. In other words, here I am interested in the role played by stock photography as ambient imagery, that is, as a mundane visual genre that permeates our everyday lives in cities, and which therefore also contributes to shaping its "moods" and "atmospheres" (Roquet 2016; Rose, Degen, and Melhuish 2014). Building on my previous work on the visual communication of specific identities in generic images (Aiello 2013; Aiello and Woodhouse 2016; Thurlow, Aiello, and Portmann, 2020), I also aim to critically examine the relationship between the aesthetic qualities of stock photography and their experiential import as ambient imagery.

Overall, the aims of this chapter are more properly conceptual rather than empirical, while also being rooted in observations collected over several years. The chapter is thus centred on a discussion of stock photography as ambient imagery informed by key academic literature on ambient media and communication and underpinned by an exploratory analysis of some key ways in which stock photography is used in outdoor public spaces, mostly in the city where I reside (Leeds) but also in other European cities where I have happened to spend some time. Before I delve into this discussion, however, in the next section I offer a focused overview of stock photography's "visual economy", that is, some of the "social relations, practices and institutions" (Rose 2010, 62) that shape its production, distribution, and circulation.

A snapshot on the visual economy of stock photography

Paul Frosh has defined the stock photography business as the "Leviathan of the image" (2013, 131), both due to its increasing importance as "an industrialized system of image-production" (Frosh 2003, 3) and its inexorable consolidation into the hands of fewer and fewer corporate giants, with Getty Images as the global market leader. Since the early 2000s, with the advent of Internet-native "microstock" agencies and the extensive digitalization of both production and distribution processes, stock photography has become ubiquitous in all arenas of media culture – from overtly corporate or commercial spheres to the more ambiguous realms of lifestyle, political, and interpersonal communication. As of 2012, the global market for stock images amounted to US$2.88 billion and was spread across 2500 commercial image suppliers mostly located in Europe, the United States and Asia (Glückler and Panitz 2013). By 2020, this market had grown to over US$4 billion and was estimated to grow by another US$1.82 billion during the period 2020-2024 (Technavio 2019). Perhaps not surprisingly, at the time of writing it has also been forecast that the "communication services" sector would see an upsurge due to the COVID-19 pandemic and register a higher growth rate compared to the global GDP growth.

The growing need for visuals as part of bite-sized and short-form digital media content developed for a variety of institutions and businesses, together with "crunching deadlines, reducing budgets, and increasing inefficiencies" (Arizton Advisory and Intelligence 2019), makes for a particularly favourable setting for the stock images market to thrive in the face of recession. Stock images are highly popular on social media platforms such as Facebook, Instagram, Pinterest, and Snapchat – something that makes their global growth prospects particularly favourable in post-pandemic times (Technavio 2019). In addition, the social power of stock photography has grown with the increasing blurring of boundaries between "pre-produced" or "ready-to-use" imagery and both editorial and commissioned photography. Stock imagery has become a staple of media, such as newspapers and advertising, that traditionally relied on "original" photography, or imagery made for the purposes of portraying a given event or promoting a specific product or service. With the growing availability of free creative images (Laurent 2014) and the gradual disappearance of exclusive licensing models to boot (Burgett 2019), the already pervasive presence of stock photographs is only likely to expand further – particularly in the realm of public space and the urban built environment.

In recent years, I and others have begun to examine the significance of generic visuals such as stock photos in news-making, specifically in relation to issues like the standardization and commodification of photojournalistic practices and styles (Aiello 2012; Runge 2020), the repurposing of a limited variety of images and visual discourses across a wide range of news stories

(Machin and Polzer 2015; Thurlow, Aiello, and Portmann 2020), the circulation of stock images across digital media platforms (Aiello 2016), and the forms of engagement that generic images may promote in relation to issues reported in the news media (Generic Visuals in the News). For the most part, however, existing scholarship on stock photography has pre-eminently focused on its role in advertising and lifestyle marketing, particularly with regards to the ways in which stock images are used to promote stereotypes (Machin 2004), authenticity (Frosh 2003; 2013), and more recently also diversity (Aiello and Woodhouse 2016).

Still visibly missing from this growing constellation of scholarship on stock imagery is an appraisal of what, in his monograph on public screens, Krajina defines as the "environmental character" of stock photographs, which also ought to be examined "as objects that form part of everyday spaces" (2014, 26-27). Not unlike television and other screen media that have become part and parcel of our everyday landscapes, stock photography's "ability to dissolve into a place's structures" (McCarthy 2001, 14) makes an investigation of its relationship with urban space particularly urgent. Most certainly, just like outdoor advertising, stock photography in cities "is a truly mass medium in an age of media fragmentation" insofar as "its occupation of urban space means that it potentially speaks to an audience that is broad in terms of demographic characteristics" (Cronin 2010, 172). At the same time, the stock photographs that we now often see on various urban surfaces are typically there to "mark" a business or at times even an otherwise vacant space. They have quite literally taken on the role of placeholders for otherwise unused or undefined spaces. In doing so, and as we will see later in this chapter, one of the main aims of this imagery may be to regulate the mood or enhance the atmosphere of its urban surroundings rather than advertise a particular product or service.

Stock photography as ambient imagery

According to Frosh, stock photography is an ambient medium because of its "absent-minded" nature, together with the fact that it's an "image-ecology" in its own right – that is, a web of relations across multiple images. In other words, from a social standpoint, stock photos exist first and foremost as an aggregate of images that we traverse "in a complex force-field of attention and distraction" (2020, 189). For this reason, Frosh argues, it becomes all the more important to examine stock photography "as a fluctuating visual environment" rather than "in relation to the power of individual images which demand (and amply repay) intense, focused viewing" (2020, 201). This is a recognition that, in turn, requires the critic to move away from "standard" discussions of the content and ideological import of particular images to focus instead on their ability to foster experiences of "aggregated, absent-minded sociality" (Frosh 2020, 189–190). Ultimately, Frosh concludes that by giving away with the "hermeneutics of suspicion" and the focus on engaged publics foregrounded by most visual culture scholars, we can start to consider "the visual promotion and proliferation of genericity as a positive social, political or ethical force" (2020, 201) and, as a consequence, stock photography as a public good rather than a "bad object".

Frosh's argument is intentionally provocative, but it also convincingly foregrounds the importance of looking at images – particularly ubiquitous imagery like stock photography – in ways that account for their *experiential* qualities as part of a broader set of social relations rather than as individualized tokens of power-laden ways of seeing. However, here I argue that this is something that can and ought to be achieved also by considering some of the key *aesthetic* qualities that set apart stock imagery. While both the *absent-mindedness* and *multiplicity* that Frosh highlights are key to the definition of stock photography as ambient imagery, I would also like to advance the idea that stock photography's *familiarity* and *flexibility* are what makes it distinctively able to "blend" into urban space and therefore also operate as part of a place's general "atmosphere". Both familiarity and flexibility are qualities that relate to stock photography's representational and design resources – that is, both what is typically included in these kinds of images and how these images are arranged to achieve a certain "effect" or "feel" (Aiello 2012). I will return to a discussion of both familiarity and flexibility in the next section.

While an emphasis on representational and design resources may be associated with the "hermeneutics of suspicion" that Frosh encourages us to leave behind, I maintain that a close attention to visual aesthetics contributes to a better understanding of stock photography as ambient imagery, in conjunction with considerations regarding its absent-mindedness and multiplicity. Not unlike outdoor advertising, and in fact often as part of branded advertisements, stock photography can be considered as what Cronin defines as "an urban, visual vernacular" (2010, 190) based on Sharon Zukin's definition of cities as "a visual repertoire of culture in the sense of a public language" (Zukin 1995, 264). As Cronin explains, the status of advertising as a vernacular originates from "the familiarity of its presence and form" (2010, 190) in conjunction with the fact that it populates the urban environment and our everyday lives in it. She then adds that advertising's status as a vernacular is also linked to its central role in relation to popular and pleasurable yet also fraught consumption practices (e.g. shopping) and commodities.

Thanks to the growing availability of inexpensive royalty-free digital images through online image banks, stock imagery is now a major if underestimated urban visual vernacular. Not only have we become accustomed to its ubiquitous presence independent of traditional advertising, but many of us are also increasingly familiar with the "typical" aesthetic of stock photography as a widespread visual genre via digital culture – many jobs nowadays require some degree of engagement with image banks for the purposes of in-house design or advertising, and social media are notoriously fertile ground for the proliferation of stock photo-based parodies and memes (Shifman 2013). As we will see in a moment, stock photographs are now regularly used to dress urban surfaces that would otherwise remain "empty" while also being amply used by businesses and institutions of all kinds to promote their services, most often in close proximity of their physical premises.

It is precisely thanks to the multiplicity of its manifestations in the world that Frosh sees stock photography as an exemplary part of a "pre-public framework" (2020, 190). However, in keeping with Zukin (1995), I would instead argue that precisely thanks to its abundance, stock photography is also and foremost a "public language", or even better a "public art" (Hariman and Lucaites 2016), albeit one that is rooted in absent-mindedness and genericity rather than focused viewing and iconicity. In other words, not unlike more arresting documentary or news imagery, stock photography may very well also provide "a way of being in the world; that is, a primary way of seeing and being seen in association with others" (Hariman and Lucaites 2020, 2). This is a view that becomes especially important in light of stock imagery's abundance in our everyday lives in cities.

Stock photography's familiar and flexible strangers in the city

Stock photography's genericity is a commercial imperative that is premised upon visual choices aimed at making the people and situations it portrays immediately recognizable as similar to those we "know" in real life while also keeping their portrayal open to a range of interpretations. In other words, both familiarity and flexibility are key to the success (or at least the usability) of stock photographs. In doing so, the "stock aesthetic" also typically relies on design resources like decontextualization (Machin 2004) and stylization

(Aiello 2013b) in order to achieve a timeless quality for images that may be used to portray "types" rather than specific objects or individuals (Aiello 2012) in broad settings such as "the home" or "the office". However, since the early 2000s, the range of design resources that set apart stock photography as a visual genre has expanded to include stylistic features that are traditionally associated with editorial, documentary or social media photography (Aiello and Woodhouse 2016). Regardless of their overarching "look", the subject matter of stock photographs is nevertheless still predominantly related to aspirational ideals of productivity and consumption. As far as representational resources are concerned, stock images tend to privilege the portrayal of individuals, couples, or small groups of people (particularly nuclear families) engaged in business, leisure, or "lifestyle" activities (Aiello and Parry 2020).

In urban public space stock photography is often used to promote businesses and services through smiling "strangers" whose identities remain undefined while also fitting in recognizable "types", to enliven otherwise vacant or drab storefronts, and ultimately also to confer warmth or texture to otherwise bland if not alienating "landscapes of capital" such as those occupied and outlined by banking services and shopping malls. I now turn to a discussion of three vignettes illustrating these uses of stock photography in urban public space.

Populating the city with perfect strangers

For over three months during the first wave of the COVID-19 pandemic, my partner, young baby, and I could only take walks around our neighbourhood in Leeds, England. On our daily outings, we often took a route that had us walk by a dental practice located in one of the typical red-brick terraced houses that line many of the city's streets. Photographs of smiling people against blank backgrounds were carefully laid out on the windows so as to cover a range of human types and relationships: two heterosexual couples (one younger and one "senior", respectively), a traditional family of four, and an attractive young woman (Figure 10.1, left). Perhaps because our lockdown-driven daily life was in stark contrast with our previous lifestyle as international academics, and more recently also as a transnational family, I found myself anticipating the comforting sight of those smiling faces, which I would have otherwise simply overlooked as a mundane part of my surroundings. Mask-less and seemingly carefree, these strangers looking out from the dental practice's windows had become noticeable in their slightly tone-deaf familiarity.

Figure 10.1. Populating the city with strangers in Leeds, UK (left) and Bern, Switzerland (right). Photos by author.

And while the exceptional circumstances underpinning our frequent walks by the dental practice made these stock images stand out in an unusual manner, many cities are populated with these "perfect strangers" that dot a variety of urban surfaces and façades. Prior to the pandemic, on my regular work trips to other European cities I often walked by stock photos of smiling individuals promoting otherwise faceless services – like, for example, the availability of a currency exchange counter at the main train station in Bern, Switzerland (Figure 10.1, right).

As Frosh explains, these generic individuals look like "approximations" of people that we have encountered before, and by virtue of their ubiquity in everyday life we have become "habituated" to "the perpetual ambient presence of strangers in both public and intimate spaces" (2020, 201). The fact that they are usually set against blank or blurred backgrounds – as in the traditional stock photo aesthetic – makes them flexible insofar as they can be used to communicate a variety of concepts and identities, and in fact we often see some of the very same subjects being used to promote different products, businesses or services. Hence their flexibility also contributes further to their increased familiarity. This said, these subjects are overwhelmingly white in appearance, though at times also ethnically

Figure 10.2. Stock photography's familiar and flexible stranger in Leeds, UK. Photo by author.

ambiguous like the young woman pictured on her own on one of the dental practice's windows. This particular model, whose name is Rebecca Givens, is also especially familiar, as for over a decade she was one of the most widely used subjects in royalty-free imagery (Figure 10.2). Overall, these images are also fairly homogenous from an emotional standpoint, and it is precisely this "upbeatness" of portrayed subjects that sets apart stock photography as a visual vernacular. Ultimately, the "perfect strangers" that we encounter in urban space through these stock images may also contribute to an overarching affect centred in the creation and regulation of a "mood" that privileges individualized approaches to identifying and engaging with the urban environment (Roquet 2016). I will return to this point in the conclusion.

Animating precarious urban façades

By virtue of the familiarity and flexibility of their subjects, stock photos are also frequently used to "animate" urban façades which would otherwise remain vacant or blank. Well before the COVID-19 pandemic, the impact of austerity had turned many a storefront in British cities into perpetually empty windows looking into dark interiors. It is now common to see these windows covered in imagery pointing to a host of potential commercial endeavours, which are unfailingly presented as being embodied in the lifestyle-oriented activities of smiling individuals — we thus routinely encounter the perfect strangers of stock photography as they are having coffee with friends, doing yoga, carrying shopping bags, or receiving beauty treatments (Figure 10.3, top). Along the same lines, stock photos can be found covering the side or unused windows of a variety of establishments, often featuring uplifting visual content that has little or nothing to do with the products or services that are sold inside (Figure 10.3, bottom). As a whole, these images conjure up a sense of imagination for these otherwise precarious spaces and surfaces.

For this reason, here it may be helpful to consider Cronin's argument that the significance of outdoor advertising may be best understood through Bergson's concept of "fabulation", which emphasizes the productive nature of advertisements as part of our immediate environment and therefore also as contributing to our embodied understanding "of the social and natural world" (2010, 126). According to this view, advertising ought to be examined as a set of stimuli that contribute to the creation of "helpful fictions that enable us to engage with the world and facilitate our actions" rather than "illusions or fictions that distort or mask true human needs" (Cronin 2010, 187). In other words, advertisements may offer practical orientations towards urban space which in turn are productive of lifelike fictions that help us make sense of the city in an embodied manner. Unlike outdoor advertising, however, the stock photos that cover unused windows and shuttered storefronts are overwhelmingly unbranded, often unanchored to text, and ostensibly open-ended insofar as they "sell" not so much a particular product or service, but rather a vision or an experience which may or may not be directly applicable to the space they occupy. These visions and experiences are abstract by virtue of these images' visual flexibility (i.e. these could be any groups of friends having coffee or any meadow with a blue sky) but are also concrete thanks to the familiarity of their visual contents (i.e. these could be my friends or a meadow and a blue sky I saw on a particular occasion). Overall, then, these are fictions that animate the surfaces that they cover by concealing their precarity and visualizing a sense that the city is rife with potential for meaningful forms of individual engagement with the surrounding environment.

Figure 10.3. Animating precarious façades in Leeds, UK (top) and Bologna, Italy (bottom). Photos by author.

Texturizing landscapes of capital

In concert with its reliance on familiar and flexible subjects and its ability to animate urban surfaces, stock photography as ambient imagery is also often set apart by a "texturized" look and feel, which is mobilized in more overtly capitalist and consumption-driven urban settings like, for example, banks and street shopping malls (Figure 10.4). In my previous work on branding and stock photography, I have defined texturization as a host of design resources, or visual cues, that are aimed at invoking the material, embodied and sensorial qualities of imagery (Aiello and Pauwels 2014). In visual communication, texture is a visual rendition of haptic or tactile features, and whether or not one can truly "touch" it, texture also summons us to identify with experiential rather than purely symbolic meaning potentials (Djonov and Van Leeuwen 2011). For example, the recent proliferation of irregular or distressed graphics that are clearly produced with computers is to be linked with the increasing importance of experiential meaning potentials like "authentic", "individual" and "personal", which have acquired particular value in an age where digital reproduction and templates are dominant (Mosbæk Johannessen and van Leeuwen 2018).

Conferring texture to stock images requires an effort to achieve greater contextualization than the typical "stock photo aesthetic" of blank or blurred backgrounds and neutral or bland colour palettes, for example through the careful staging of cluttered, layered or grainy settings and surfaces. At the same time, these images can be texturized through lighting and photographic effects (Aiello and Woodhouse 2016). In Figure 10.4, top, we see what looks like a blown-up stock rendition of a social media image in the window of a bank located on the corner of two busy commercial streets in the city centre of Leeds. An uplifting scene of heterosexual ordinary romance set in a park is enhanced by the "warmth" of its filter and sun flare, a carefully chosen photographic "glitch". The anchoring text reveals the bank's aim to "humanize" its customers and, in this way, make us engage with its services as individuals. Along the same lines, in Figure 10.4, bottom, the large side windows of a supermarket in a street shopping mall of a Danish city are covered in texturized images of spring onions and the hands and forearm of a man kneading dough on a table top covered in flour. The chiaroscuro lighting brings out the grain of both the spring onions and the arms and hands working on the dough. Not unlike images found in the business class menus examined by Crispin Thurlow in his critique of privilege in airline travel, here "a rich sense of photographic texture is produced through the surfaces depicted, as well as with the organic materiality of the vegetables"

Figure 10.4. Texturizing landscapes of capital in Leeds, UK (top) and Nyborg, Denmark (bottom). Photos by author.

(2020, 11), something that in turn produces a sense of distinction for those who get to choose their meals from these exclusive menus.

As a whole, staged images that make texture "visible" bring us closer, in an experiential and embodied manner, to both the visual text at hand and the urban site in which this is placed. Drawing from Sedgwick (2003), Thurlow states that "touch is unavoidably also that other kind of feeling: that feel-good feeling and that "feeling good about myself feeling" (2020, 15). In texturizing these urban landscapes of capital through the familiar and flexible textures of stock images like the ones I have just described, the impersonal and often ruthless structures of finance and commerce may make us feel close to them and may also make us feel good on an individual, personal level.

Conclusion: ambient imagery and the "mood" of the city

Just like the air we breathe, ambient imagery is both pervasive and unremarkable, and as mundane as it is part and parcel of our everyday lives. In particular, sourced from millions of possibilities in online image banks, stock photography's banality and its abundance in urban space make it both strikingly insignificant and integral to our ways of being and moving in and through the city. Here, I have outlined some of the key ways in which stock photography is mobilized in urban space, particularly with regards to the relationship between some of its experiential and aesthetic dimensions. In linking stock photography's visual familiarity and flexibility with how we may encounter and experience it in the city, I have attempted to start thinking through some of the ways in which, as Thurlow (2020) points out, the "sensory qualities of texts" may also contribute to "their sensuous, affective resonances" (2020, 14-15). I would therefore now like to conclude this chapter with a brief reflection on how the interaction of stock photography's key aesthetic qualities – i.e. familiarity and flexibility – with the urban built environment contributes to creating the "mood" of everyday life in cities.

In his book on ambient media in Japan, Paul Roquet traces the relationship between the emergence of "mood-regulating" media and a neoliberal culture of therapy and healing that relies on "atmospheric determinations of self" (2016, 15). As he states: "The demand for self-care has shifted not just media use but media *aesthetics*" (Roquet 2016, 16), where "the imperative for calming affect" and for "providing a sense of restfulness and relaxation" (Roquet 2016, 18) has become central to music, film, video art, and even literature. Ambient media provide what he considers as provisional comfort in the face of the threats and uncertainties of life under capitalism, while also functioning as a form of social control. This said, according to Roquet, ambient media also offer affordances for publics to reflect on urban coexistence and participate in collective practices of attunement. Along the same lines, stock photography's ambient qualities are centred on aesthetic choices that may make urban sites and surfaces that are often unremarkable and utilitarian if not downright precarious and alienating more engaging and even comforting. This is a mood that feeds on encounters with familiar and flexible subjects which may make us feel close to these sites and surfaces, and also good about ourselves as we traverse the city. In doing so, however, these encounters also promote individualized – or dare we say, neoliberal – approaches to living with others in cities. At the same time, this ambient imagery personally invites us into the spaces it occupies in ways that may enable us to start imagining what life in the city could be like in a more sensuous, embodied manner.

References

Aiello, Giorgia. 2007. "The Appearance of Diversity: Visual Design and the Public Communication of EU Identity." In *European Union Identity: Perceptions from Asia and Europe*, edited by Jessica Bain and Martin Holland, 147-181. Baden-Baden: Nomos.

———. 2012. "The 'Other' Europeans: the Semiotic Imperative of Style in *Euro Visions* by Magnum Photos." *Visual Communication* 11 (1): 49-77.

———. 2013a. "Generiche Differenze: La Comunicazione Visiva Della Soggettività Lesbica Nell'archivio Fotografico Getty Images." *Studi Culturali* anno X n. 3: 523-548.

———. 2013b. "Fra Abiezione e Stilizzazione: Corpi Femminili, Corpi Lesbici e Corpi Queer Nella Comunicazione Visiva Globale." *AG About Gender – Rivista Internazionale di Studi di Genere* 2 (3): 145-163.

———. 2016. "Taking Stock." Ethnography Matters. http://ethnographymatters.net/blog/2016/04/28/taking-stock/

Aiello, Giorgia, and Anna Woodhouse. 2016. "When Corporations Come to Define the Visual Politics of Gender: The Case of Getty Images." *Journal of Language and Politics* 15 (3): 352–368.

Aiello, Giorgia, and Katy Parry. 2020. *Visual Communication: Understanding Images in Media Culture*. London: SAGE.

Arizton Advisory and Intelligence. 2019. "Stock Images and Video Markets: Global Outlook and Forecast 2019-2024." https://www.arizton.com/market-reports/stock-images-and-videos-market-report-2024/.

Burgett, Gannon. 2019, Nov 7. "Getty Images is Phasing Out Rights-Managed 'Creative Images' in Favor of Royalty Free Content." Digital Photography Review. https://www.dpreview.com/news/0475276807/getty-is-phasing-out-rights-managed-photos-in-favor-of-royalty-free-content/.

Cronin, Anne. 2010. *Advertising, Commercial Spaces and the Urban*. Basingstoke: Palgrave Macmillan.

Djonov, Emilia, and Theo van Leeuwen. 2011. "The Semiotics of Texture: From Tactile to Visual." *Visual Communication* 10 (4): 541–564.

Frosh, Paul. 2003. *The Image Factory: Consumer Culture, Photography and the Visual Content Industry*. Oxford: Berg.

———. 2013. "Beyond the Image Bank: Digital Commercial Photography." In *The Photographic Image in Digital Culture*, edited by Martin Lister, 131–148. London: Routledge.

———. 2020. "Is Commercial Photography a Public Evil? Beyond the Critique of Stock Photography." In *Photography and its Publics*, edited by Melissa Miles and Ed Welch, 187-206. London: Bloomsbury.

Generic Visuals in the News. 2021. "Generic Visuals in the News: The Role of Stock Photos and Simple Data Visualizations in Assembling Publics." University of Leeds. https://genericvisuals.leeds.ac.uk/.

Gendelman, Irina, and Giorgia Aiello. 2010. "Faces of Places: Façades as Global Communication in Post-Eastern Bloc Urban Renewal." In *Semiotic Landscapes: Language, Image, Space*, edited by Adam Jaworski and Crispin Thurlow, 256–273. London: Continuum.

Glückler, Johannes, and Robert Panitz R. 2013. "Survey of the Global Stock Image Market 2012." GSIM Research Group. http://www.stockimagemarket.uni-hd.de/.

Gürsel, Zeynep. 2016. *Image Brokers: Visualizing World News in the Age of Digital Circulation*. Berkeley, CA: University of California Press.

Hariman, Robert, and John Lucaites. 2016. *The Public Image: Photography and Civic Spectatorship*. Chicago: University of Chicago Press.

Hariman, Robert, and John Lucaites. 2020. "Photography and Public Culture." In *Photography and its Publics*, edited by Melissa Miles and Ed Welch, 21-37. London: Bloomsbury.

Krajina, Zlatan. 2014. *Negotiating the Mediated City: Everyday Encounters with Public Screens*. New York: Routledge.

Laurent, Olivier. 2014, March 5. "Getty Images Makes 35 Million Images Free to Use." *British Journal of Photography*. https://www.bjp-online.com/2014/03/gettyimages-makes-35-million-images-free-in-fight-against-copyright-infringement/.

Machin, David. 2004. "Building the World's Visual Language: The Increasing Global Importance of Image Banks in Corporate Media." *Visual Communication* 3, no. 3: 316–336.

Machin, David, and Lydia Polzer. 2015. *Visual Journalism*. London: Palgrave Macmillan.
McCarthy, Anna. 2001. *Ambient Television: Visual Culture and Public Space*. Durham and London: Duke University Press.
Mosbæk Johannessen, Christian, and Theo van Leeuwen. 2018. "(Ir)regularity." In *The Materiality of Writing: A Trace-Making Perspective*, edited by Christian Mosbæk Johannessen and Theo van Leeuwen, 175-192. London: Routledge.
Roquet, Paul. 2016. *Ambient Media: Japanese Atmospheres of the Self*. Minneapolis, MN: University of Minnesota Press.
Rose, Gillian. 2010. *Doing Family Photography: The Domestic, the Public and the Politics of Sentiment*. Farnham: Ashgate.
Rose, Gillian, Degen, Monica, and Clare Melhuish. 2014. "Networks, Interfaces, and Computer-Generated Images: Learning from Digital Visualisations Of Urban Redevelopment Projects." *Environment and Planning D: Society and Space* 32, no. 3: 386–403.
Runge, Evelyn. 2020. "The Travels Of Photographs Within the Global Image Market: How Monopolisation, Interconnectedness, And Differentiation Shape the Economics Of Photography." *Photographies* 13 (3): 365-384.
Sedgwick, Eve Kosofsky. 2003. *Touching Feeling: Affect, Pedagogy, Performativity*. Duke University Press, Durham, NC.
Shifman, Limor. 2013. *Memes in Digital Culture*. Cambridge, MA: The MIT Press.
Technavio. 2019. "Stock Images Market by Application, Image Source, License Model, and Geography – Forecast and Analysis 2020-2024." https://www.technavio.com/report/stock-images-market-industry-analysis/.
Thurlow, Crispin. 2020. "Touching Taste: The Classist Intersections of Materiality, Textuality And Tactility." *Multimodal Communication* 9 (2): 1-16.
Thurlow, Crispin, Aiello, Giorgia, and Lara Portmann. 2020. "Visualizing Teens and Technology: A Social Semiotic Analysis of Stock Photography and News Media Imagery." *New Media & Society* 22 (3): 528-549.
Zukin, Sharon. 1995. *The Cultures of Cities*. Malden: Blackwell.

About the author

Giorgia Aiello is Professor of Culture and Communication at the University of Leeds (UK) and Associate Professor in Sociology of Culture and Communication at the University of Bologna (Italy). She has written extensively about the problems and potentials of visual communication in digital media, cities, and everyday life. She is a co-author of *Visual Communication: Understanding Images in Media Culture* (with Katy Parry; SAGE 2020) and a co-editor of *Communicating the City: Meanings, Practices, Interactions* (with Matteo Tarantino and Kate Oakley; Peter Lang, 2017). Her current research projects focus on the role of generic images in contemporary culture, stock photography and data visualization in digital journalism, and visual-material approaches to researching the urban built environment.

List of Works Cited

Abram, Simone, and Gisa Weszkalnys. 2011. "Introduction: Anthropologies of Planning—Temporality, Imagination, and Ethnography." *Focaal* 61: 3-18.

Adam, Barbara. 2006. "Futurescapes: Challenge for Social and Management Science." Paper presented at the 5th Palermo International Conference on Social Time Retroscapes and Futurescapes, Palazzo d' Aumale, Terrasini, June 2006. https://citeseerx.ist.psu.edu/viewdoc/download?doi=10.1.1.101.1105&rep=rep1&type=pdf/.

Adur, Shweta M., and Shreyasi Jha. 2018. "(Re)Centering Street Harassment – An Appraisal of Safe Cities Global Initiative in Delhi, India." *Journal of Gender Studies* 27 (1): 114–24.

Adut, Ari. 2012. "A Theory of the Public Sphere." *Sociological Theory* 30 (4): 238-262.

Agostinho, Daniela. 2018. "Chroma Key Dreams: Algorithmic Visibility, Fleshy Images and Scenes of Recognition." *Philosophy of Photography* 9 (2): 131–55.

Agre, Phil. 1994. "Surveillance and Capture: Two Models of Privacy." *The Information Society* 10 (2): 101–27.

Ahad, Annie Dayani, and Syamimi Md Ariff Lim. 2014. "Convenience or Nuisance?: The 'WhatsApp' Dilemma." *Procedia — Social and Behavioral Sciences* 155: 189–96.

Aiello, Giorgia. 2007. "The Appearance of Diversity: Visual Design and the Public Communication of EU Identity." In *European Union Identity: Perceptions from Asia and Europe*, edited by Jessica Bain and Martin Holland, 147–181. Baden-Baden: Nomos.

———. 2012. "The 'Other' Europeans: the Semiotic Imperative of Style in *Euro Visions* by Magnum Photos." *Visual Communication* 11 (1): 49–77.

———. 2013a. "Generiche Differenze: La Comunicazione Visiva Della Soggettività Lesbica Nell'archivio Fotografico Getty Images." *Studi Culturali* anno X n. 3: 523–548.

———. 2013b. "Fra Abiezione e Stilizzazione: Corpi Femminili, Corpi Lesbici e Corpi Queer Nella Comunicazione Visiva Globale." *AG About Gender – Rivista Internazionale di Studi di Genere* 2 (3): 145–163.

———. 2016. "Taking Stock." Ethnography Matters. http://ethnographymatters.net/blog/2016/04/28/taking-stock/

Aiello, Giorgia, and Anna Woodhouse. 2016. "When Corporations Come to Define the Visual Politics of Gender: the Case of Getty Images." *Journal of Language and Politics* 15 (3): 352–368.

Aiello, Giorgia, and Katy Parry. 2020. *Visual Communication: Understanding Images in Media Culture*. London: SAGE.

Akrich, M. 1992. "The De-scription of Technological Objects." In *Shaping Technology/ Building Society: Studies in Sociotechnical Change*, edited by Wiebe E. Bijker and John Law, 205–224. Cambridge, MA: MIT Press.

Allen, Louisa. 2011. "'Picture This': Using Photo-Methods in Research on Sexualities and Schooling." *Qualitative Research* 11 (5): 487–504.

Amin, Ash. 2008. "Collective Culture and Urban Public Space." *City* 12 (1): 5–24.

Amin, Ash and Thrift, Nigel. 2002. *Cities: Reimagining the Urban*. Cambridge: Polity.

Amoore, Louise. 2018. "Cloud Geographies: Computing, Data, Sovereignty." *Progress in Human Geography* 42 (1): 4–24.

———. 2020. *Cloud Ethics: Algorithms and the Attributes of Ourselves and Others*. Durham, NC: Duke University Press.

Anderson, Steve F. 2017. *Technologies of Vision: The War Between Data and Images*. Cambridge, MA: MIT Press.

Andrejevic, Mark, and Mark Burdon. 2014. "Defining the Sensor Society." *Television & New Media* 16 (1): 19–36.

Anthopoulos, Leonidas. 2017. "Smart Utopia vs Smart Reality: Learning by Experience from 10 Smart City Cases." *Cities* 63 (March): 128–48.

Appadurai, Arjun. 1995. "The Production of Locality." In *Counterworks: Managing the Diversity of Knowledge*, edited by R. Fardon, 208–29. London: Routledge.

Aravot, Iris. 2002. "Back to Phenomenological Placemaking." *Journal of Urban Design* 7 (2): 201–212.

Archambault, Julie Soleil. 2011. "Breaking up 'Because of the Phone' and the Transformative Potential of Information in Southern Mozambique." *New Media & Society* 13 (3): 444–56.

———. 2017. *Mobile Secrets: Youth, Intimacy, and the Politics of Pretence in Mozambique*. Chicago: University of Chicago Press.

Arizton Advisory and Intelligence. 2019. "Stock Images and Video Markets: Global Outlook and Forecast 2019-2024." https://www.arizton.com/market-reports/stock-images-and-videos-market-report-2024/.

Ash, James. 2015. *The Interface Envelope: Gaming, Technology, Power*. New York: Bloomsbury.

———. 2018. *Phase Media: Space, Time and the Politics of Smart Objects*. London: Bloomsbury.

Ash, James, Ben Anderson, Rachel Gordon, and Paul Langley. 2017. "Unit, Vibration, Tone: A Post-Phenomenological Method for Researching Digital Interfaces." *Cultural Geographies* 25 (1): 165–81.

Ashburner, Xanthe. 2019. "Fabrizio Terranova (Dir.). Donna Haraway: Storytelling for Earthly Survival. Icarus Films, 2016." *Australian Feminist Studies* 34 (99): 128–130.

Auger, James. 2013. "Speculative Design: Crafting the Speculation." *Digital Creativity*, 24 (1): 11–35.

Aurigi, Alessandro, and Katherine Willis, eds. 2020. *The Routledge Companion to Smart Cities*. Abingdon: Routledge.

Awad, Edmond, Sohan Dsouza, Richard Kim, Jonathan Schulz, Joseph Henrich, Azim Shariff, Jean-François Bonnefon, and Iyad Rahwan. 2018. "The Moral Machine Experiment." *Nature* 563: 59–64.

Axelrod, Robert. 1973. "Schema Theory: An Information Processing Model of Perception and Cognition." *American Political Science Review* 67 (4): 1248–1266.

Back, Les, Shamser Sinha, and Charlynne Bryan. 2012. "New Hierarchies of Belonging." *European Journal of Cultural Studies*. 15 (2): 139–154.

Ballatore, Andrea, Mark Graham, and Shilad Sen. 2017. "Digital Hegemonies: The Localness of Search Engine Results." *Annals of the American Association of Geographers* 107 (5): 1194–1215.

Barad, K. 2007. *Meeting the Universe Halfway: Quantum Physics and the Entanglement of Matter and Meaning*. Durham, NC: Duke University Press.

———. 2014. "Diffracting Diffraction: Cutting Together-Apart." *Parallax* 20 (3): 168–87.

———. 2015. "TransMaterialities: Trans*/Matter/Realities and Queer Political Imaginings." *GLQ: A Journal of Lesbian and Gay Studies* 21 (2–3): 387–422.

Barns, Sarah. 2019. "Negotiating the Platform Pivot: From Participatory Digital Ecosystems to Infrastructures of Everyday Life." *Geography Compass* 13 (9): https://doi.org/10.1111/gec3.12464.

———. 2020a. *Platform Urbanism: Negotiating Platform Ecosystems in Connected Cities*. Basingstoke: Palgrave Macmilllan.

———. 2020b. "Tools of the Trade: Urban Planning, Urban Media and the Refashioning of Urban Space." In *The Routledge Companion to Urban Media and Communication*, edited by Zlatan Krajina and Deborah Stevenson, 236–47. Abingdon: Routledge.

Barthes, R. 1973. "Theory of the Text." In *Untying the Text: A Post-Structuralist Reader*, edited by Robert Young, translated by Ian McLeod, 31–48, Boston: Routledge, 1981.

Basu, Paul, and Sharon Macdonald. 2007. "Introduction: Experiments in Exhibition, Ethnography, Art, and Science." In *Exhibition Experiments*, edited by Sharon Macdonald and Paul Basu, 1–24. Oxford: Blackwell.

Bate, David. 2013. "The Digital Condition of Photography: Cameras, Computers and Display." In *The Photographic Image in Digital Culture*, edited by Martin Lister, 91–108. London: Routledge.

Batty, M. 1990. "Intelligent Cities: Using Information Networks to Gain Competitive Advantage." *Environment and Planning B: Planning and Design* 17 (3): 247–56.

BBC. 2017. "Museum of London Gets £180 Million Funding Boost." BBC News, January 24, 2017, https://www.bbc.com/news/uk-england-london-38732919?intlink_from_url=&/.

Bell, Vikki. 2016. "Between Documentality and Imagination: Five Theses on Curating the Violent Past." *Memory Studies* 11 (2): 137–55.

Bendon, Helen. 2005. "A Place for Ambiguity: Articulating Practice as Research." *Journal of Media Practice* 6 (3): 157–65.

Benjamin, Ruha. 2019. *Race After Technology: Abolitionist Tools for the New Jim Code.* Cambridge: Polity Press.

Benjamin, Walter. 1999. *The Arcades Project,* Cambridge, MA: Harvard University Press.

Bennett, Jane. 2010. *Vibrant Matter: A Political Ecology of Things.* Durham: Duke University Press.

Berger, John. 1972. *Ways of Seeing.* London: BBC with Penguin.

Besse, Jean-Marc. 2013. "European Cities from a Bird's-Eye View: The Case of Alfred Guesdon." In *Seeing From Above: The Aerial View in Visual Culture,* edited by Mark Dorrian and Frédéric Pousin, 66–82. London: IB Tauris.

Bhan, Gautam, and Kalyani Menon-Sen. 2008. *Swept off the Map: Surviving Eviction and Resettlement in Delhi.* New Delhi: Yoda Press.

Bingham-Hall, John, and Stephen Law. 2015. "Connected or Informed?: Local Twitter Networking in a London Neighbourhood." *Big Data & Society* 2 (2): https://doi.org/10.1177/2053951715597457.

Binns, Amy. 2014. "Twitter City and Facebook Village: Teenage Girls' Personas and Experiences Influenced by Choice Architecture in Social Networking Sites." *Journal of Media Practice* 15 (2): 71–91.

Bishop, Christopher M. 2006. *Pattern Recognition and Machine Learning.* New York: Springer.

Bissell, David. 2009. "Travelling Vulnerabilities: Mobile Timespaces of Quiescence." *Cultural Geographies,* October.

Bivens, Rena. 2018. "Exploiting a Dystopic Future to Unsettle Our Present-Day Thinking About Sexual Violence Prevention." *Ada New Media* (blog). 10 May 2018. https://adanewmedia.org/2018/05/issue13-bivens/.

Bleecker, Julian, and Nicolas Nova. 2009. "A Synchronicity: Design Fictions for Asynchronous Urban Computing." In *Situated Technologies Pamphlets,* edited by Omar Khan, Trebor Scholz, and Mark Shepard. Vol. 5. New York: The Architectural League.

Bordo, Susan. 1993. *Unbearable Weight: Feminism, Western Culture, and The Body.* Berkeley: University of California Press.

Bowles Eagle, Ryan. 2015. "Loitering, Lingering, Hashtagging: Women Reclaiming Public Space Via #BoardtheBus, #StopStreetHarassment, and the #EverydaySexism Project." *Feminist Media Studies* 15 (2): 350–53. https://doi.org/10.1080/14680777.2015.1008748/.

Boy, John D, and Justus Uitermark. 2017. "Reassembling the City through Instagram." *Transactions of the Institute of British Geographers* 42 (4): 612–24.

Boyer, Marie-Christine. 1994. *The City of Collective Memory: Its Historical Imagery and Architectural Entertainments*. London: MIT Press.

———. 1996. *CyberCities: Visual Perception in the Age of Electronic Communication*. New York: Princeton Architectural Press.

Bratton, Benjamin. 2015. *The Stack: On Software and Sovereignty*. Cambridge, MA: MIT Press.

Braudel, Fernand. 1984. *The Perspective of the World*. New York: Harper & Rowe.

Braun, Erik, Mihalis Kavaratzis, and Sebastian Zenker. 2013. "My City – My Brand: the Different Roles of Residents in Place Branding." *Journal of Place Management and Development*, 6 (1): 18-28.

Bridge, Gary and Watson, Sophie. 2003. "City Publics." In *A Companion to the City*, edited by Gary Bridge and Sophie Watson, 369–379. Oxford: Blackwell.

Brown, Barry, and Eric Laurier. 2017. "The Trouble With Autopilots: Assisted and Autonomous Driving on the Social Road." *CHI '17: Proceedings of the 2017 CHI Conference on Human Factors in Computing Systems* May: 416–29. https://doi.org/10.1145/3025453.3025462/.

Browne, Simone. 2015. *Dark Matters: On the Surveillance of Blackness*. London: Duke University Press.

Budge, Kylie. 2020. "Visually Imagining Place: Museum Visitors, Instagram, and the City." *Journal of Urban Technology*. 27 (2): 61–79.

Bunz, Mercedes. 2019. "The Calculation Of Meaning: On the Misunderstanding Of New Artificial Intelligence As Culture." *Culture, Theory and Critique* 60 (3-4): 264-78.

Burgett, Gannon. 2019, Nov 7. "Getty Images is Phasing Out Rights-Managed 'Creative Images' in Favor of Royalty Free Content." Digital Photography Review. https://www.dpreview.com/news/0475276807/getty-is-phasing-out-rights-managed-photos-in-favor-of-royalty-free-content/.

Butcher, Melissa. 2018. "Defying Delhi's Enclosures: Strategies for Managing a Difficult City." *Gender, Place & Culture* 25 (5): 727–42.

Butler, Judith. 2007. "Torture and the Ethics of Photography." *Environment and Planning D: Society and Space* 25 (6): 951–66.

Candy, Linda, and Ernest Edmonds. 2018. "Practice-Based Research in the Creative Arts: Foundations and Futures from the Front Line." *Leonardo* 51 (1): 63–69.

Cardullo, Paolo, and Rob Kitchin. 2019. "Smart Urbanism and Smart Citizenship: The Neoliberal Logic of 'Citizen-Focused' Smart Cities in Europe." *Environment and Planning C: Politics and Space* 37 (5): 813–30.

Carmody, Pádraig. 2013. "A Knowledge Economy or an Information Society in Africa? Thintegration and the Mobile Phone Revolution." *Information Technology for Development* 19 (1): 24–39.

Casetti, Francesco. 2015. *The Lumière Galaxy: Seven Key Words for the Cinema to Come*. New York: Columbia University Press.

Castells, Manuel. 1989. *The Information City: Information Technology, Economic Restructuring, and The Urban-Regional Process*. Oxford: Blackwell.

Chamaa, Sabine El. 2017. "Mind the Gap! Working Papers on Practice-Based Doctoral Research in the Creative Arts and Media." *Journal of Media Practice* 18 (1): 83–85.

Chant, Sylvia. 2013. "Cities through a 'Gender Lens': A Golden 'Urban Age' for Women in the Global South?" *Environment and Urbanization* 25 (1): 9–29. https://doi.org/10.1177/0956247813477809.

Chapman, Owen, and Kim Sawchuk. 2012. "Creation: Intervention, Analysis and 'Family Resemblances'." *Canadian Journal of Communication* 37 (1): 5–26.

Cheney-Lippold, John. 2017. *We Are Data: Algorithms and the Making of Our Digital Selves*. New York: New York University Press.

Chung, Hye Jean. 2018. *Media Heterotopias: Digital Effeccts and Material Labour in Global Production*. London: Duke University Press.

City of London. 2018. "Cultural Strategy 2018-2022." City of London Corporation. https://www.cityoflondon.gov.uk/assets/Things-to-do/city-of-london-cultural-strategy-2018-22.pdf/.

Clark, Catherine E. 2018. *Paris and The Cliché of History: The City and Photographs, 1860 to 1970*. Oxford: Oxford University Press.

Clough, Patricia Ticineto. 2018. *The User Unconscious: Affect, Media and Measure*. London: University of Minnesota Press.

Coleman, Rebecca. 2018. "Social Media and the Materialisation of the Affective Present." In *Affect and Social Media: Emotion, Mediation, Anxiety and Contagion*, edited by Tony D. Sampson, Stephen Maddison, and Darren Ellis, 67–75. London: Rowman & Littlefield International.

Coleman, Rebecca, and Richard Tutton. 2017. "Introduction to Special Issue of Sociological Review on 'Futures in Question: Theories, Methods, Practices'." *The Sociological Review* 65 (3): 440-447.

Copier, Laura, and Eliza Steinbock. 2018. "On Not Really Being There: Trans* Presence/Absence in Dallas Buyers Club." *Feminist Media Studies* 18 (5): 923–41.

Corbridge, Stuart. 2005. "Governance." In *Seeing the State: Governance and Governmentality in India*, edited by Stuart Corbridge, Glyn Williams, M. K. Srivastava, and Rene Veron, 151–87. Cambridge: Cambridge University Press.

Cordts, M., Mohamed Omran, Sebastian Ramos, Timo Rehfeld, Markus Enzweiler, Rodrigo Benenson, Uwe Franke, Stefan Roth, and Bernt Schiele. 2016. "The Cityscapes Dataset for Semantic Urban Scene Understanding." In *Proceedings of the IEEE Conference on Computer Vision and Pattern Recognition (CVPR)*.

Cowley, Robert, Simon Joss, and Youri Dayot. 2018. "The Smart City and Its Publics: Insights from across Six UK Cities." *Urban Research & Practice* 11 (1): 53–77.

Crang, Michael, Tracey Crosbie, and Stephen Graham. 2006. "Variable Geometries of Connection: Urban Digital Divides and the Uses of Information Technology." *Urban Studies* 43 (13): 2551–70.

———. 2007. "Technology, Time-space, and the Remediation of Neighbourhood Life." *Environment and Planning A* 39: 2405–2422.

Crary, Jonathan. 1990. *Techniques of the Observer: On Vision and Modernity in the Nineteenth Century*. Cambridge, Mass: MIT Press.

Crawford, Kate and Trevor Paglen. 2019. "Excavating AI: The Politics of Images in Machine Learning Training Sets." https://www.excavating.ai/.

Cresswell, Tim. 1996. *In Place/Out of Place: Geography, Ideology, and Transgression*. Minneapolis: University of Minnesota Press.

Cronin, Anne M. 2010. *Advertising, Commercial Spaces and the Urban*. Basingstoke: Palgrave Macmillan.

Cronin, Anne M., and Kevin Hetherington, eds. 2008. *Consuming the Entrepreneurial City: Image, Memory, Spectacle*. London and New York: Routledge.

Cultural Hub Working Party. 2013. "Cultural Hub Working Party Report to Policy and Resources Committee." City of London Corporation. https://democracy.cityoflondon.gov.uk/ieIssueDetails.aspx?IId=35627&Opt=3/.

Culture Mile. 2020a. "Between the Storeys." Culture Mile. https://www.culturemile.london/betweenthestoreys/betweenthestoreys/.

———. 2020b. Culture Mile. https://www.culturemile.london/.

Datta, Ayona. 2012. *The Illegal City: Space, Law and Gender in a Delhi Squatter Settlement*. Farnham: Ashgate.

———. 2018. "The 'Digital Turn' in Postcolonial Urbanism: Smart Citizenship in the Making of India's 100 Smart Cities." *Transactions of the Institute of British Geographers* 43 (3): 405–419.

Datta, Ayona, and Arya Thomas. 2021. "Curating #AanaJaana [#ComingGoing]: Gendered Authorship in the "Contact Zone" of Delhi's Digital and Urban Margins." Cultural Geographies, https://doi.org/10.1177/1474474021993415/.

Davies, Alex. 2018. "How Do Self-Driving Cars See? (And How Do They See Me?)." *Wired*, August 11, 2018.

Davies, David. 2004. *Art As Performance*. Malden, MA: Blackwell.

Degen, Monica Montserrat. 2008. *Sensing Cities: Regenerating Public Life in Barcelona and Manchester*. London: Routledge.

———. 2010. "Consuming Urban Rhythms: Let's Ravalejar." In *Geographies of Rhythm: Nature, Place, Mobilities and Bodies*, edited by Tim Edensor, 21–32, Aldershot: Ashgate.

———. 2014. "The Everyday City of the Senses." In *Cities and Social Change: Encounters with Contemporary Urbanism*, edited by Eugene McCann, and Ronan Paddison, 92–111, London: SAGE.

———. 2018. “Timescapes of Urban Change: The Temporalities of Regenerated Streets.” *The Sociological Review* 66 (5): 1074–1092.

Degen, Monica, Clare Melhuish, and Gillian Rose. 2017. “Producing Place Atmospheres Digitally: Architecture, Digital Visualisation Practices and the Experience Economy.” *Journal of Consumer Culture* 17 (1): 3–24.

Degen, Monica, and Gillian Rose. 2022. *The New Urban Aesthetic: Digital Experiences of Urban Change*. London: Bloomsbury Academic.

Deleuze, Gilles. 1992. “Postscript on the Societies of Control.” *October* 59: 3–7.

Della Lucia, Maria, and Mariapina Trunfio. 2018. “The Role of the Private Actor in Cultural Regeneration: Hybridizing Cultural Heritage with Creativity in the City.” *Cities* 82: 35–44.

Denson, Shane. 2016. “Crazy Cameras, Discorrelated Images, and the Post-Perceptual Mediation of Post-Cinematic Affect.” In *Post-Cinema: Theorizing 21st-Century Film*, edited by Shane Denson and Julia Leyda, 193–233. Falmer: Reframe.

———. 2020. *Discorrelated Images*. Durham NC: Duke University Press.

Denson, Shane, and Julia Leyda, eds. 2016. *Post-Cinema: Theorizing 21st-Century Film*. Falmer: Reframe. http://reframe.sussex.ac.uk/post-cinema/.

Dixon, Natalie. 2018. “Stranger-Ness and Belonging in a Neighbourhood WhatsApp Group.” *Open Cultural Studies* 1 (1): 493–503.

Djonov, Emilia, and Theo van Leeuwen. 2011. “The Semiotics of Texture: From Tactile to Visual.” *Visual Communication* 10 (4): 541–564.

Donald, James. 1999. *Imagining the Modern City*. London: Athlone Press.

Dreyfus, Hubert L. 1991. *Being-in-the-World: A Commentary on Heidegger’s Being and Time, Division I*. Cambridge, MA: MIT Press.

Dunne, Anthony, and Fiona Raby. n.d. “SpeculativeEdu | Dunne & Raby: Designers Need to Recognise Their Limitations and Work with Them.” Accessed February 8, 2021. https://speculativeedu.eu/interview-dunne-raby/.

Duru, Asli. 2018. “Wearable Cameras, in-Visible Breasts: Intimate Spatialities of Feminist Research with Wearable Camcorders in Istanbul.” *Gender, Place & Culture* 25 (7): 939–54.

———. 2019. “‘A Walk down the Shore’: A Visual Geography of Ordinary Violence in Istanbul.” *Environment and Planning D. Society and Space* 37(6): 1064–1080.

———. 2020. “A Walk Down the Shore| Sahilde 1 Gezinti.” *Cultural Geographies* 27 (1): 157–61.

Edensor, Tim. 2001. “Performing Tourism, Staging Tourism: (Re)producing Tourist Space and Practice.” *Tourist Studies* 1 (1): 59–81.

Edwards, Elizabeth, ed. *1992. Anthropology and Photography 1860-1920*. London.

Elsaesser, Thomas. 2013a. “Digital Cinema: Convergence or Contradiction?” In *The Oxford Handbook of Sound and Image in Digital Media*, edited by Carol Vernallis, Amy Herzog, and John Richardson, 12–44. Oxford: Oxford University Press.

———. 2013b. "The 'Return' of 3-D: On Some of the Logics and Genealogies of the Image in the Twenty-First Century." *Critical Inquiry* 39 (2): 217–46.

Elwood, Sarah. 2020. "Digital Geographies, Feminist Relationality, Black and Queer Code Studies: Thriving Otherwise." *Progress in Human Geography*, January, 030913251989973.

Elwood, Sarah, and Agnieszka Leszczynski. 2018. "Feminist Digital Geographies." Gender, Place & Culture 25 (5): 629–44.

Erickson, Kristin. 2020. "The Promises and Pitfalls Of Cyber Urbanism: Governance And Participation." In *The Routledge Companion to Urban Media and Communication,* edited by Zlatan Krajina and Deborah Stevenson, 225–235. London and New York: Routledge.

Eubanks, Virginia. 2017. *Automating Inequality: How High-Tech Tools Profile, Police, and Punish the Poor.* New York, NY: St Martin's Press.

Evans, Leighton. 2015. *Locative Social Media*. London: Palgrave Macmillan.

Farrell, Terry. 2007. *Smithfield-Farringdon: Reinforcing Urban Identity & Character.* London: English Heritage.

Featherstone, Mike. 1998. "The 'Flâneur', the City and Virtual Public Life." *Urban Studies* 35 (5/6): 909–925.

Flint, Kate. 2000. *The Victorians and the Visual Imagination*. Cambridge: Cambridge University Press.

Florida, Richard. 2005. *Cities and the Creative Class*. New York: Routledge.

Flusser, Vilem. 2011. *Into the Universe of Technical Images*. Minneapolis: University of Minnesota Press.

Forensic Architecture. 2018. "Synthetic Data Generation: Development of Data Classification Tools." https://synthetic-datasets.sfo2.digitaloceanspaces.com/website/FA_Synthetic-Data-Generation_Report.pdf/.

Friend, John and Allen Hickling. 1997. *Planning Under Pressure: A Strategic Choice Approach.* 2nd edition. London: Butterworth-Heinemann.

Frith, Jordan. 2015. *Smartphones as Locative Media*. Cambridge: Polity.

Frosh, Paul. 2003. *The Image Factory: Consumer Culture, Photography and the Visual Content Industry.* Oxford: Berg.

———. 2013. "Beyond the Image Bank: Digital Commercial Photography." In *The Photographic Image in Digital Culture*, edited by Martin Lister, 131–148. London: Routledge.

———. 2020. "Is Commercial Photography a Public Evil? Beyond the Critique of Stock Photography." In *Photography and its Publics*, edited by Melissa Miles and Ed Welch, 187–206. London: Bloomsbury.

Fukushima, Kunihiko. 1980. "Neocognitron: A Self-organizing Neural Network Model for a Mechanism of Pattern Recognition Unaffected by Shift in Position." *Biological Cybernetics* 36 (4): 193–202.

Gabrys, Jennifer. 2014. "Programming Environments: Environmentality and Citizen Sensing in the Smart City." *Environment and Planning D: Society and Space* 32 (1): 30–48.

———. 2019. "Sensors and Sensing Practices: Reworking Experience Across Entities, Environments, and Technologies." *Science, Technology, & Human Values* 44 (5): 723-36.

Gabrys, Jennifer, and Helen Pritchard. 2018. "Sensing Practices." In *Posthuman Glossary*, edited by Rosi Braidotti, and Maria Hlavajova, 394-96. London: Bloomsbury.

Ganesh, Maya Indira. 2017. "Entanglement: Machine Learning and Human Ethics In Driver-less Car Crashes." *A Peer-Reviewed Journal About* (*APRJA*) 6 (1): 76-87.

Gekker, Alex and Sam Hind. 2019. "Infrastructural Surveillance." *New Media & Society* 22 (8): 1414–36.

Gendelman, Irina, and Giorgia Aiello. 2010. "Faces of Places: Façades as Global Communication in Post-Eastern Bloc Urban Renewal." In *Semiotic Landscapes: Language, Image, Space*, edited by Adam Jaworski and Crispin Thurlow, 256–273. London: Continuum.

Generic Visuals in the News. 2021. "Generic Visuals in the News: The Role of Stock Photos and Simple Data Visualizations in Assembling Publics." University of Leeds. https://genericvisuals.leeds.ac.uk/.

Germanidis, A., and Cristóbal Valenzuela. n.d. Uncanny Rd (website). Accessed March 23, 2021. https://uncannyroad.com/.

Gershgorn, Dave. 2017. "The Data That Transformed AI Research — And Possibly the World." Quartz. https://qz.com/1034972/the-data-that-changed-the-direction-of-ai-research-and-possibly-the-world/.

Giannini, Tula, and Jonathan P Bowen. 2016. "Curating Digital Life and Culture: Art and Information. "In *Electronic Visualisation and the Arts* (*EVA*). http://dx.doi.org/10.14236/ewic/EVA2016.46/.

Gillespie, Tarleton. 2010. "The Politics of 'Platforms'." *New Media & Society* 12 (3): 347–64.

Gitlin, Todd. 2001. *Media Unlimited: How the Torrent of Images and Sounds Overwhelms Our Lives*. New York: Metropolitan Books.

Glückler, Johannes, and Robert Panitz R. 2013. "Survey of the Global Stock Image Market 2012." GSIM Research Group. http://www.stockimagemarket.uni-hd.de/.

Goodfellow, Ian, Jean Pouget-Abadie, Mehdi Mirza, Bing Xu, David Warde-Farley, Sherjil Ozair, Aaron Courville, and Yoshua Bengio. 2014. "General Adversarial Nets." In *NIPS'14: Proceedings of the 27th International Conference on Neural Information Processing Systems — Volume 2*, edited by Z. Ghahramani, M. Welling, C. Cortes, N. Lawrence, and K. Q. Weinberger, 2672–2680. https://papers.nips.cc/paper/2014/hash/5ca3e9b122f61f8f06494c97b1afccf3-Abstract.html/.

Goodfellow, Ian, Yoshua Bengio and Aaron Courville. 2016. *Deep Learning*. Cambridge, MA: MIT Press.

Gordon, Eric. 2010. *The Urban Spectator: American Concept Cities from Kodak to Google*. Hanover, NH: Dartmouth College Press.

Gordon, Eric, and Adriana de Souza e Silva. 2011. *Net Locality: Why Location Matters in a Networked World*. Oxford: Wiley-Blackwell.

Graham, Stephen D. N. 2005. "Software-Sorted Geographies." *Progress in Human Geography* 29 (5): 562–80.

Greenberg, Miriam. 2000. "Branding Cities: A Social History of the Urban Lifestyle Magazine." *Urban Affairs Review* 36 (2): 228–263.

Greenfield, Adam. 2013. *Against the Smart City*. London: Do Projects.

Gürsel, Zeynep. 2016. *Image Brokers: Visualizing World News in the Age of Digital Circulation*. Berkeley, CA: University of California Press.

Gurumurthy, Anita, Nandini Chami, and Sanjana Thomas. 2016. "Unpacking Digital India: A Feminist Commentary on Policy Agendas in the Digital Moment." *Journal of Information Policy* 6: 371–402.

Habermas, Jürgen. 1991. *Structural Transformation of the Public Sphere: An Inquiry into a Category of Bourgeois Society*. Cambridge, MA: MIT Press.

———. 1996. *Between Facts and Norms: Contributions to a Discourse Theory of Law and Democracy*. Cambridge, MA: MIT Press.

Hackers of Resistance. 2018. "hord14r13z" *Ada: A Journal of Gender, New Media, and Technology* 13, https://dx.doi.org/10.5399/uo/ada.2018.13.6/.

Halegoua, Germaine R. 2019. *The Digital City: Media and the Social Production of Place*. New York: New York University Press.

Halpern, Orit. 2015. *Beautiful Data: A History of Vision and Reason since 1945*. Durham NC: Duke University Press.

Hand, Martin. 2012. *Ubiquitous Photography*. Cambridge: Polity Press.

———. 2020. "Photography Meets Social Media: Image Making and Sharing in a Continually Networked Present." In *The Handbook of Photography Studies*, edited by Gil Pasternak, 310–326. London: Bloomsbury.

Hand, Martin, and Ashley Scarlett. 2019. "Habitual Photography: Time, Rhythm, and Temporalization in Contemporary Personal Photography." In *The Routledge Companion to Photography Theory*, edited by Mark Durden and Jane Tormey, 410–25. New York: Routledge.

Hansen, Mark. 2009. "Living (with) Technical Time: From Media Surrogacy to Distributed Cognition." *Theory, Culture & Society* 26 (2-3): 294–315.

———. 2015. *Feed-Forward: On the Future of Twenty-First-Century Media*. Chicago: University of Chicago Press.

Haraway, Donna. 2016. *Staying with the Trouble: Making Kin in the Chthulucene*. Durham, NC: Duke University Press.

Harding, Sandra. 1991. *Whose Science? Whose Knowledge?: Thinking from Women's Lives.* Ithaca: Cornell University Press.

Hariman, Robert, and John Lucaites. 2016. *The Public Image: Photography and Civic Spectatorship.* Chicago: University of Chicago Press.

———. 2020. "Photography and Public Culture." In *Photography and its Publics*, edited by Melissa Miles and Ed Welch, 21–37.

Harman, Graham. 2002. *Tool-Being: Heidegger and the Metaphyics of Objects.* Chicago: Open Court.

Harper, Graeme. 2008. "Creative Writing: Words as Practice-Led Research." *Journal of Visual Art Practice* 7 (2): 161–71.

Harvey, Adam. 2019. "MegaPixels: Face Recognition Training Datasets." https://ahprojects.com/megapixels/.

Harvey, David. 1989. "From Managerialism To Entrepreneurialism: The Transformation In Urban Governance in Late Capitalism." *Geografiska Annaler: Series B, Human Geography* 71 (1): 3–17.

Heeks, Richard. 2008. "ICT4D 2.0: The Next Phase of Applying ICT for International Development." *Computer* 41 (6): 26–33.

Heidegger, Martin. 1962. *Being and Time.* Oxford: Blackwell.

Helmond, Anne. 2015. "The Platformization of the Web: Making Web Data Platform Ready." *Social Media + Society* 1 (2): https://doi.org/10.1177/2056305115603080.

Hesse-Biber, Sharlene Nagy, and Amy J. Griffin. 2015. "Feminist Approaches to Multimethod and Mixed Methods Research." In *The Oxford Handbook of Multimethod and Mixed Methods Research Inquiry*, edited by Sharlene Nagy Hesse-Biber and R. Burke Johnson, 72–89, Oxford: Oxford University Press.

Highfield, Tim. 2016. *Social Media and Everyday Politics.* Cambridge: Polity.

Hind, Sam. 2019. "Digital Navigation and the Driving-Machine: Supervision, Calculation, Optimization, And Recognition." *Mobilities* 14 (4): 401–17.

Hjorth, Larissa, and Sarah Pink. 2014. "New Visualities and the Digital Wayfarer: Reconceptualizing Camera Phone Photography and Locative Media." *Mobile Media & Communication* 2 (1): 40–57.

Hoch, Charles. 2000. "Planning Craft: How Planners Compose Plans." *Planning Theory* 8 (3): 219–241.

Hochman, Nadav. 2014. "The Social Media Image." *Big Data & Society* 1 (2): https://doi.org/10.1177/2053951714546645.

———. 2017. "From Site-Specificity to Hyper-Locality: Performances of Place in Social Media." In *The SAGE Handbook of Social Media Research Methods*, edited by Luke Sloan and Anabel Quan-Haase, 367–85. London: Sage.

Hochman, Nadav, and Lev Manovich. 2013. "Zooming into an Instagram City: Reading the Local through Social Media." *First Monday*, June: https://doi.org/10.5210/fm.v18i7.4711.

Hodson, Mike, Julia Kasmire, Andrew McKeekin, John G Stehlin, and Kevin Ward, eds. 2020. *Urban Platforms and the Future City: Transformations in Infrastructure, Governance, Knowledge and Everyday Life*. Abingdon: Routledge.

Hoelzl, Ingrid, and Rémi Marie. 2015. *Softimage: Towards a New Theory of the Digital Image*. Bristol: Intellect Books.

———. 2016. "Brave New City: the Image of the Urban Data-space." *Visual Communication* 15(3): 371–391.

Hollands, Robert G. 2008. "Will the Real Smart City Please Stand Up?" *City* 12 (3): 303–20.

———. 2016. "Beyond the Corporate Smart City?: Glimpses of Other Possibilities of Smartness." In *Smart Urbanism: Utopian Vision or False Dawn*, edited by Simon Marvin, Andrés Luque-Ayala, and Colin McFarlane, 169–85. London: Routledge.

Hong, Sun-ha. 2016. "Data's Intimacy: Machinic sensibility and the Quantified Self." *Communication +1* 5 (3): 1–36.

hooks, bell. 1992. *Black Looks: Race and Representation*. Boston: South End Press.

Hubbard, Phil, and Tim Hall. 1998. "The Entrepreneurial City and the 'New Urban Politics'." In *The Entrepreneurial City: Geographies of Politics, Regime and Representation*, edited by Tim Hall and Phil Hubbard, 1–23, Chicester: Wiley.

Hubel, David H. and Torsten Wiesel. 1959. "Receptive Fields of Single Neurones in the Cat's Striate Cortex." *Journal of Physiology* 148 (3): 574–91.

Hui, Yuk. 2016. *On the Existence of Digital Objects*. Minneapolis: University of Minnesota Press.

Huyssen, Andreas. 2008. "Introduction: World Culture, World Cities." In *Other Cities, Other Worlds: Urban Imaginaries in a Globalizing Age*, edited by Andreas Huyssen, 1–23. Durham: Duke University Press.

Imarisha, Walida. 2015. "Introduction." In *Octavia's Brood: Science Fiction Stories from Social Justice Movements*, edited by Adrienne Maree Brown and Walida Imarisha, Edinburgh: AK Press.

Ingold, Tim. 2004. "Culture on the Ground: World Perceived Through the Feet." *Journal of Material Culture* 9 (3): 315–340.

Isin, Engin, and Evelyn Ruppert. 2015. *Being Digital Citizens*. London: Rowman & Littlefield.

Jacobs, Jane. 1961. *The Death and Life of Great American Cities*. Harmondsworth: Pelican.

Jensen, Ole. 2020. "On the Move: on Mobile Agoras, Networked Selves, and the Contemporary City." In *The Routledge Companion to Urban Media and Communication*, edited by Zlatan Krajina and Deborah Stevenson, 96–106. London and New York: Routledge.

Jirón, Paola A, Walter Alejandro Imilan, and Luis Iturra. 2016. "Relearning to Travel in Santiago: The Importance of Mobile Place-Making and Travelling Know-How." *Cultural Geographies* 23 (4): 599–614.

Johansson, Lotta. 2016. "Post-Qualitative Line of Flight and the Confabulative Conversation: A Methodological Ethnography." *International Journal of Qualitative Studies in Education* 29 (4): 445–66.

John, Nicholas. 2016. *The Age of Sharing*. Cambridge: Polity.

Jukes, Stephen. 2018. "The Newsroom Is No Longer a Safe Zone: Assessing the Affective Impact of Graphic User-Generated Images on Journalists Working with Social Media." In *Affect and Social Media: Emotion, Mediation, Anxiety and Contagion*, edited by Tony D. Sampson, Stephen Maddison, and Darren Ellis, 168–77. London: Rowman & Littlefield International.

Kabeer, Naila. 1999. "Resources, Agency, Achievements: Reflections on the Measurement of Women's Empowerment." *Development and Change* 30 (3): 435–64.

Kamath, Anant. 2018. "'Untouchable' Cellphones? Old Caste Exclusions and New Digital Divides in Peri-Urban Bangalore." *Critical Asian Studies* 50 (3): 375–94.

Karvonen, Andrew, Federico Cugurullo, and Federico Caprotti, eds. 2018. *Inside Smart Cities: Place, Politics and Urban Innovation*. 1st ed. London: Routledge.

Katz, J.E., and M. Aakhus. 2002. *Perpetual Contact: Mobile Communication, Private Talk, Public Performance*. Cambridge: Cambridge University Press.

Kaun, Anne, and Fredrik Stiernstedt. 2014. "Facebook Time: Technological and Institutional Affordances for Media Memories." *New Media & Society* 16 (7): 1154–68.

Keightley, Emily, and Philip Schlesinger. 2014. "Digital Media – Social Memory: Remembering in Digitally Networked Times." *Media, Culture & Society* 36 (6): 745–47.

Keller, Jessalynn, Kaitlynn Mendes, and Jessica Ringrose. 2018. "Speaking 'Unspeakable Things': Documenting Digital Feminist Responses to Rape Culture." *Journal of Gender Studies* 27 (1): 22–36.

Kember, Sarah, and Joanna Zylinska. 2012. *Life After New Media: Mediation as a Vital Process*. Cambridge, MA: MIT Press.

Kennedy, Maev. 2016. "Off to the Market: Museum of London Shows Off its New Smithfield Site." The Guardian, May 20, 2016.

Kenyon, Nicholas. 2018. "The Culture Mile that Will Transform Arts in the City." Evening Standard, 16 March, 2018. https://www.standard.co.uk/comment/comment/the-culture-mile-that-will-transform-arts-in-the-city-a3792026.html/.

Kitchin, Rob. 2014. "The Real-Time City? Big Data and Smart Urbanism." *GeoJournal* 79 (1): 1–14.

Kitchin, Rob, and Martin Dodge. 2011. *Code/Space: Software and Everyday Life*. London: MIT Press.

Kitchin, Rob, Claudio Coletta, and Gavin McArdle. 2017. "Urban Informatics, Governmentality and the Logics of Urban Control." The Programmable City Working Paper 25. http://progcity.maynoothuniversity.ie/.

Kitchin, Rob, Sophia Maalsen, and Gavin McArdle. 2016. "The Praxis and Politics of Building Urban Dashboards." *Geoforum* 77 (December): 93–101.

Kleine, Dorothea, and Sammia Poveda. 2016. "Digital Divide." In *International Encyclopedia of Geography: People, the Earth, Environment and Technology*, edited by Douglas Richardson, Noel Castree, Michael F. Goodchild, Audrey Kobayashi, Weidong Liu, and Richard A. Marston, 1–5, London: John Wiley & Sons.

Klingmann, Anna. 2007. *Brandscapes: Architecture in the Experience Economy.* Cambridge, MA: MIT Press.

Konstantinidou, Eirini. 2017. "Practicing Theory: Mnemophrenia – a Film-Essay on the Future of Cinema and Artificial Memories." *Journal of Media Practice* 18 (1): 4–13.

Krajina, Zlatan. 2014. *Negotiating the Mediated City: Everyday Encounters with Public Screens.* London and New York: Routledge.

———. 2016. "The Alternative Urbanism of Psychogeography in the Mediated City." In *Radical Space: Exploring Politics and Practice*, edited by Debra Shaw and Maggie Humm, 39-63. London: Rowman & Littlefield.

———. 2019. "Understanding Encounters for Urban Media Studies: Civic Intercourse, Screen Technologies, and Cultural Difference." *International Journal of Communication* 13: 5351–5369.

Kristensen, Dorthe Brogård, and Minna Ruckenstein. 2018. "Co-Evolving with Self-Tracking Technologies." *New Media & Society* 20 (10): 3624–40.

Krizhevsky, Alex, Ilya Sutskever and Geoffrey Hinton. 2012. "ImageNet Classification with Deep Convolutional Neural Networks." In *NIPS'12: Proceedings of the 25th International Conference on Neural Information Processing Systems – Volume 1.* https://papers.nips.cc/paper/2012/file/c399862d3b9d6b76c8436e924a68c45b-Paper.pdf/.

Lane, Joshua. 2016. "The Digital Street: an Ethnographic Study of Networked Street Life in Harlem." *American Behavioural Scientist* 60 (1): 43–58.

Lather, Patti, and Elizabeth A. St Pierre. 2013. "Post-Qualitative Research." *International Journal of Qualitative Studies in Education* 26 (6): 629–33.

Laurent, Olivier. 2014, March 5. "Getty Images Makes 35 Million Images Free to Use." *British Journal of Photography.* https://www.1854.photography/2014/03/getty-images-makes-35-million-images-free-in-fight-against-copyright-infringement/#:~:text=The%20stock%20agency's%20controversial%20move,from%20new%20mobile%20photography%20players./.

Lazzarato, Maurizio. 2019. *Videophilosophy: The Perception of Time in Post-Fordism.* Translated by Jay Hetrick. New York: Columbia University Press.

LeCun, Yann, Patrick Haffner, Léon Bottou and Yoshua Bengio. 1999. "Object Recognition with Gradient-Based Learning." In *Shape, Contour and Grouping in Computer Vision,* edited by David A. Forsyth, Joseph. L. Mundy, Vito di Gesú, and Roberto Cipolla, 319–345. Berlin: Springer.

Leorke, Dale, and Christopher Wood. 2019. "'Alternative Ways of Being': Reimagining Locative Media Materiality through Speculative Fiction and Design." *Media Theory* 3 (2): 63–102.

Lessig, Lawrence. 2009. "Remix: Making Art and Commerce Thrive in the Hybrid Economy." http://archive.org/details/LawrenceLessigRemix/.

Leszczynski, Agnieszka. 2019a. "Platform Affects of Geolocation." *Geoforum* 107: 207–15.

———. 2019b. "Glitchy Vignettes of Platform Urbanism." *Environment and Planning D: Society and Space,* October, 026377581987872.

Levin, Sam. 2020. "Safety Driver Charged In 2018 Incident Where Self-Driving Uber Car Killed a Woman." *The Guardian*, September 16, 2020.

Levitt, Deborah. 2018. *The Animatic Apparatus: Animation, Vitality, and the Futures of the Image*. Winchester: Zero Books.

Lew, Alan A. "Tourism Planning and Place Making: Place-Making or Placemaking?" *Tourism Geographies* 19 (3): 448–466.

Lindner, Christoph, and Miriam Meissner, eds. 2019. *The Routledge Companion to Urban Imaginaries*. Abingdon: Routledge.

Lloyd, Justine. 2013. "Trouble Comes to Me: the Mediated Place of the Urban Citizen." *Space and Culture* 16 (3): 306-322.

Loukissas, Yanni A. 2019. *All Data Are Local: Thinking Critically in a Data-Driven Society*. Cambridge, MA: MIT Press.

Lovink, Geert, and Ned Rossiter. 2019. "Propositions on the Organizational Form." In *Organize*, edited by Timon Beyes, Lisa Conrad, and Reinhold Martin, 89-101. Lüneberg: Meson Press.

Lucarelli, Andrea. "Place Branding as Urban Policy: The (Im)political Place Branding." *Cities* 80: 12-21.

Lupton, Deborah. 2016. *The Quantified Self: A Sociology of Self-Tracking*. Cambridge: Polity Press.

———. 2020. *Data Selves*. Cambridge: Polity Press.

Luque-Ayala, Andrés, and Simon Marvin. 2016. "The Maintenance of Urban Circulation: An Operational Logic of Infrastructural Control." *Environment and Planning D: Society and Space* 34 (2): 191–208.

Lyotard, J-F. 1984. *The Postmodern Condition* (1979). Manchester: Manchester University Press.

Machin, David. 2004. "Building the World's Visual Language: the Increasing Global Importance of Image Banks in Corporate Media." *Visual Communication* 3 (3): 316–336.

Machin, David, and Lydia Polzer. 2015. *Visual Journalism*. London: Palgrave Macmillan.

Mackenzie, Adrian. 2017. *Machine Learners: Archaeology of a Data Practice*. Cambridge, MA: MIT Press.

MacKenzie, Adrian, and Anna Munster. 2019. "Platform Seeing: Image Ensembles and Their Invisualities." *Theory, Culture & Society* 36 (5): 3–22.
Malevé, Nicholas. 2019. "An Introduction to Image Datasets." Unthinking Photography. https://unthinking.photography/articles/an-introduction-to-image-datasets
Malinowska, Anna, and Toby Miller. 2017. "Sensitive Media." *Open Cultural Studies* 1 (1). https://doi.org/10.1515/culture-2017-0060.
Manovich, Lev. 2006. "The Poetics of Augmented Space." *Visual Communication* 5 (2): 219–40.
Manovich, Lev. 2016. "What Is Digital Cinema?" In *Post-Cinema: Theorizing 21st-Century Film*, edited by Shane Denson and Julia Leyda, 20–50. Falmer: Reframe. http://reframe.sussex.ac.uk/post-cinema/.
Marchart, Oliver. 2004. "Politics and Artistic Practice: On the Aesthetics of the Public Sphere." *Frakcija* 33/34: 14–19.
Margetts, Helen, Peter John, Scott A. Hale, and Taha Yasseri. 2017. *Political Turbulence: How Social Media Shape Collective Action*. Princeton: Princeton University Press.
Marvin, Simon, and Andrés Luque-Ayala. 2017. "Urban Operating Systems: Diagramming the City." *International Journal of Urban and Regional Research* 41 (1): 84–103.
Marvin, Simon, Andres Luque-Ayala, and Colin McFarlane, eds. 2016. *Smart Urbanism: Utopian Vision or False Dawn?* Abingdon: Routledge.
Masuda, Jeffrey R., and Sonia Bookman. 2018. "Neighbourhood Branding and the Right To The City." *Progress in Human Geography* 42 (2): 165–182.
Mattern, Shannon. 2015. "Mission Control: A History of the Urban Dashboard." *Places Journal*. 2015. https://placesjournal.org/article/mission-control-a-history-of-the-urban-dashboard/.
———. 2017. "Mapping's Intelligent Agents." *Places Journal*. https://placesjournal.org/article/mappings-intelligent-agents/.
McCarthy, Anna. 2001. *Ambient Television: Visual Culture and Public Space*. Durham, NC: Duke University Press.
McCorduck, Pamela. 2004. *Machines Who Think: A Personal Inquiry Into the History and Prospects of Artificial Intelligence*. Natick: A K Peters.
McCormack, Derek. 2017. "The Circumstances of Post-Phenomenological Life Worlds." *Transactions of the Institute of British Geographers* 42: 2–13.
McCosker, Anthony, and Rowan Wilken. 2020. *Automating Vision: The Social Impact of the New Camera Consciousness*. New York: Routledge.
McCulloch, Warren S. and Walter Pitts. 1943. "A Logical Calculus of the Ideas Immanent in Nervous Activity." *Bulletin of Mathematical Biophysics* 5: 115–133.
McDermott, Fiona. 2019. "New Sensorial Vehicles: Navigating Critical Understandings of Autonomous Futures." In *Architecture and the Smart City*, edited by Sergio M. Figueiredo, Sukanya Krishnamurthy, and Torsten Schroeder, 247–251. London: Routledge.

McFarlane, Colin. 2016. "The Geographies of Urban Density: Topology, Politics and the City." *Progress in Human Geography* 40 (5): 629–48.

McKim, Joel. 2017. "Speculative Animation: Digital Projections of Urban Past and Future." *Animation* 12 (3): 287–305.

———. 2019. "Trevor Paglen Trains His Sights on the Rise of Machine Vision." Apollo Magazine. https://www.apollo-magazine.com/trevor-paglen-machine-vision/

McQuire, Scott. 2010. "Rethinking Media Events: Large Screens, Public Space Broadcasting and Beyond." *New Media and Society* 12 (4): 567–582

———. 2016. *Geomedia: Networked Cities and the Future of Public Space*. Cambridge: Polity Press.

———. 2020. "An Archeology of the Media City: Toward a Critical Cultural History of Mediated Urbanism." In *The Routledge Companion to Urban Media and Communication*, edited by Zlatan Krajina and Deborah Stevenson, 13–22. Abingdon: Routledge.

Melitopoulos, Angela. 1999. "Passing Drama by Angela Melitopoulos, Excerpt, 4, 36." YouTube video, https://www.youtube.com/watch?v=1pAqy4Uoxjg.

Merrifield, Andy. 2012. "The Politics of the Encounter and the Urbanization of the World." *City* 16 (3): 269–283.

Metz, Cade. 2018. "How Driverless Cars See the World Around Them." *The New York Times*, March 19, 2018. https://www.nytimes.com/2018/03/19/technology/how-driverless-cars-work.html/.

Miles, Malcolm, and Steven Miles. 2004. *Consuming Cities*. London: Palgrave MacmIllan.

Minsky, Marvin and Seymour A. Papert. 1969. *Perceptrons: An Introduction to Computational Geometry*. Cambridge, MA: MIT Press.

Mitchell, W. J. T. 1992. *The Reconfigured Eye: Visual Truth in the Post-Photographic Era*. Cambridge MA: MIT Press.

Mitchell, William J. 2003. *Me++: The Cyborg Self and the Networked City*. Cambridge, MA: MIT Press.

Moores, Shaun. 2014. "Digital Orientations: 'Ways of the Hand' and Practical Knowing in Media Uses and Other Manual Activities." *Mobile Media & Communication* 2 (2): 196–208.

Morley, David. 2000. *Home Territories: Media, Mobility, and Identity*. London: Routledge.

———. 2017. *Communications and Mobility: the Migrant, the Mobile Phone, and the Container Box*. London: Wiley-Blackwell.

Mosbæk Johannessen, Christian, and Theo van Leeuwen. 2018. "(Ir)regularity." In *The Materiality of Writing: A Trace-Making Perspective*, edited by Christian Mosbæk Johannessen and Theo van Leeuwen, 175–192. London: Routledge.

Mouffe, Chantal. 2008. *Art as Public Issue: How Art and Its Institutions Reinvent the Public Dimension*. Rotterdam: NAi Publishers.

Mumford, Lewis. 1937. "What is a City?" *Architectural Record* 82: 52–68.

Munster, Anna. 2006. *Materializing New Media: Embodiment in Information Aesthetics*. 1st ed. Hanover, NH: Dartmouth College Press, University Press of New England.

———. 2013. *An Aesthesia of Networks: Conjunctive Experience in Art and Technology*. London: MIT Press.

Munster, Anna, and Adrian MacKenzie. 2019. "Platform Seeing: Image Ensembles and Their Invisualities." *Theory, Culture & Society* 36 (5): 3–22.

Munster, Anna, and Geert Lovink. 2005. "Theses on Distributed Aesthetics: Or, What a Network is Not." *The Fibreculture Journal* 7. http://seven.fibreculturejournal.org/fcj-040-theses-on-distributed-aesthetics-or-what-a-network-is-not/.

Murphy, Ian, and Saint John Walker. 2019. "'Make It More Invisible': Hollywood VFX Training Crosstalk on the Transatlantic Fiber Optic." *Animation Studies* 15. https://journal.animationstudies.org/ian-murphy-saint-john-walker-make-it-more-invisible-hollywood-vfx-training-crosstalk-on-the-transatlantic-fiber-optic/.

Museum of London (MoL). 2021. Museum of London. https://www.museumoflondon.org.uk/museum-london/.

Myers, Dowell, and Alicia Kitsuse. 2000. "Constructing the Future in Planning: A Survey of Theories and Tools." *Journal of Planning Education and Research* 19 (3): 221–231.

Nancy, Jean-Luc. 2009. "Un Art de la Ville" *Caliban: French Journal of English Studies* 25: https://journals.openedition.org/caliban/1427.

National Transportation Safety Board (NTSB). 2019a. "Highway investigation HWY18MH010: Vehicle Automation Report." *NTSB*, May 11, 2019. https://data.ntsb.gov/Docket/Document/docBLOB?ID=40477717&FileExtension=.PDF&FileName=Vehicle%20Automation%20Report-Master.PDF.

———. 2019b. "Highway Investigation HWY18MH010: Human Performance Factual Report." *NTSB*, May 11, 2019. https://data.ntsb.gov/Docket/Document/docBLOB?ID=40477743&FileExtension=.PDF&FileName=Human%20Performance%20Factual-Master.PDF.

———. 2019c. "Highway investigation HWY18MH010: Uber ATG Vehicle Operator Interview." *NTSB*, May 11, 2019. https://data.ntsb.gov/Docket/Document/docBLOB?ID=40477745&FileExtension=.PDF&FileName=Human%20Performance%20Attachment%20-%20Uber%20ATG%20Vehicle%20Operator%20Interview-Master.PDF.

———. 2019d. "Highway Investigation HWY18MH010: Collision Between Vehicle Controlled By Developmental Driving System and Pedestrian, Tempe, Arizona, March 18, 2018." *NTSB*, May 11, 2019. https://data.ntsb.gov/Docket/Document/docBLOB?ID=40479021&FileExtension=.PDF&FileName=NTSB%20-%20Adopted%20Board%20Report%20HAR-19%2F03-Master.PDF.

Neff, Gina, and Dawn Nafus. 2016. *Self-Tracking*. Cambridge MA: MIT Press.

Nielsen, Michael. 2019. "Using Neural Nets to Recognize Handwriting." Neural Networks and Deep Learning. http://neuralnetworksanddeeplearning.com/chap1.html.

Niquille, Simone C. 2019. "Regarding the Pain of SpotMini: Or What a Robot's Struggle to Learn Reveals about the Built Environment." In *Machine Landscapes: Architectures of the Post-Anthropocene*, edited by Liam Young, 84–91. Oxford: Architectural Design.

Noble, Safiya Umoja. 2018. *Algorithms of Oppression: How Search Engines Reinforce Racism*. New York: New York University Press.

O'Hara, Kenton P., Michael Massimi, Richard Harper, Simon Rubens, and Jessica Morris. 2014. "Everyday Dwelling with WhatsApp." In *Proceedings of the 17th ACM Conference on Computer Supported Cooperative Work & Social Computing*, 1131–43. CSCW '14. New York: ACM. https://doi.org/10.1145/2531602.2531679/.

Olazaran, Mikel. 1996. "A Sociological Study of the Official History of the Perceptrons Controversy." *Social Studies of Science* 26 (3): 611–659.

Otter, Chris. 2008. *The Victorian Eye: A Political History of Light and Vision in Britain, 1800-1910*. London: University of Chicago Press.

Oxbotica (@Oxbotica). 2019. "Our Eyes are a Vital Asset when it Comes to Spotting and Reacting to Danger on the Roads." *Twitter*, November 8, 2019. https://twitter.com/oxbotica/status/1192815056926457858.

Oxman, E. 2010. "Sensing the Image: Roland Barthes and the Affect of the Visual." *SubStance* 39 (2): 71–90.

Packer, Jeremy, and Stephen B. Crofts Wiley. 2012. "Introduction: The Materiality of Communication." In *Communication Matters: Materialist Approaches to Media, Mobility and Networks*, edited by Jeremy Packer and Stephen B. Crofts Wiley, 3–16. London: Routledge.

Paez, Danny. 2019. "'This Person Does Not Exist' Creator Reveals His Site's Creepy Origin Story." Inverse. https://www.inverse.com/article/53414-this-person-does-not-exist-creator-interview/.

Papastergiadis, Nikos, Scott McQuire, Xin Gu, Amelia Barikin, Ross Gibson, Audrey Yue, Sun Jung, Cecelia Cmielewski, Soh Yeong Roh, and Matt Jones. 2013. "Mega Screens for Mega Cities." *Theory, Culture & Society* 30 (7-8): 325–341.

Parks, Lisa, and Nicole Starosielski. 2015. "Introduction." In *Signal Traffic: Critical Studies of Media Infrastructures*, edited by Lisa Parks and Nicole Starosielski, 1–27. Urbana, Chicago: University of Illinois Press.

Pasquale, Frank. 2017. "From Territorial to Functional Sovereignty." *Law and Political Economy Blog*, June 12, 2017. https://lpeblog.org/2017/12/06/from-territorial-to-functional-sovereignty-the-case-of-amazon/

Payne, Will. 2017. "Welcome to the Polygon: Contested Digital Neighborhoods and Spatialized Segregation on Nextdoor." *Computational Culture* 6.

Peake, Linda. 2016. "On Feminism and Feminist Allies in Knowledge Production in Urban Geography." *Urban Geography* 37 (6): 830–38.

Pelc, Noelany, Nadia Hasan, and Debra Mollen. 2020. "Feminist Storytelling: Representing the Stories of Diverse Women in Psychology." *Women & Therapy* 43 (1–2): 7–17.

Pentzold, Christian, and Manuel Menke. 2020. "Conceptualizing the Doings and Sayings of Media Practices: Expressive Performance, Communicative Understanding and Epistemic Discourse." *International Journal of Communication* 14: 2789–2809.

Phadke, Shilpa, Sameera Khan, and Shilpa Ranade. 2011. *Why Loiter?: Women and Risk on Mumbai Streets*. New Delhi: Penguin Books.

Pickford, James. 2017. "Nicholas Kenyon on London's New 'Culture Mile'." Financial Times, July 20, 2017. https://www.ft.com/content/ab89e83e-6ca3-11e7-b9c7-15af748b60d0/.

Pine, Joseph B., and James H. Gilmore. 1999. *The Experience Economy: Work Is Theatre and Every Business a Stage*. Boston: Harvard Business School.

Pinney, Christopher, and Nicolas Peterson, eds. 2003. *Photography's Other Histories*. London: Duke University Press.

Poggiali, Lisa. 2017. "Digital Futures and Analogue Pasts? Citizenship and Ethnicity in Techno-Utopian Kenya." *Africa* 87 (2): 253–77.

Pollio, Andrea. 2019. "Forefronts of the Sharing Economy: Uber in Cape Town." *International Journal of Urban and Regional Research* 43 (4): 760–75.

Prakash, Gyan. 2008. "Introduction." In *The Spaces of the Modern City: Imaginaries, Politics, and Everyday Life*, edited by Gyan Prakash and Kevin Michael Kruse, 1–18. Oxford: Princeton University Press.

Pred, Allan. 1986. *Place, Practice and Structure: Social and Spatial Transformation in Southern Sweden: 1750-1850*. Cambridge: Polity Press.

Publica. 2015. "Barbican and Golden Lane Area Strategy." August 23, 2015. https://issuu.com/publicaassociates/docs/150826_181206_barbican_strategy_final/

Punathambekar, Aswin, and Sriram Mohan. 2019. *Global Digital Cultures: Perspectives from South Asia*. Ann Arbor: University of Michigan Press.

Qiu, J. 2009. *Working-Class Network Society: Communication Technology and the Information Have-Less in Urban China*. Cambridge, MA: MIT Press.

Rangaswamy, Nimmi. 2013. "Digital Life of the Youth in India – What Is the Current Scenario and Where Are We Headed?" Digital Equality (blog). 6 December 2013. https://digitalequality.in/internet-magic-in-indian-slums/.

Rangaswamy, Nimmi, and Payal Arora. 2015. "The Mobile Internet in the Wild and Every Day: Digital Leisure in the Slums of Urban India." *International Journal of Cultural Studies* 19 (6): 611–26.

Rao, Siriginidi Subba. 2005. "Bridging Digital Divide: Efforts in India." *Telematics and Informatics*, 22 (4): 361–75.

Relph, E. C. 1976. *Place and Placelessness*. London: Pion Ltd.

Rettberg, Jill Walker. 2014. *Seeing Ourselves Through Technology: How We Use Selfies, Blogs and Wearable Devices to See and Shape Ourselves.* Basingstoke: Palgrave Macmillan.

Richards, Greg and Palmer, Robert. 2010. *Eventful Cities: Cultural Management and Urban Revitalization.* Oxford: Butterworth-Heinemann.

Richardson, Mark. 2012. "The Disintegration Loops." *Pitchfork*, November 19, 2019. https://pitchfork.com/reviews/albums/17064-the-disintegration-loops/.

Richardson, Lizzie. 2020. "Platforms, Markets, and Contingent Calculation: The Flexible Arrangement of the Delivered Meal." *Antipode* 52 (3): 619–36.

Ristola, Jacqueline. 2017. "Realist Film Theory and Flowers of Evil: Exploring the Philosophical Possibilities of Rotoscoped Animation." *Animation Studies* 12. https://journal.animationstudies.org/category/maureen-furniss-award-2017/jacqueline-ristola-realist-film-theory-and-flowers-of-evil-exploring-the-philosophical-possibilities-of-rotoscoped-animation-winner/.

Rius Ulldemolins, Joaqium. 2014. "Culture and Authenticity in Urban Regeneration Processes: Place Branding in Central Barcelona." *Urban Studies* 51 (14): 3026–3045.

Robins, Kevin. 1996. *Into the Image: Culture and Politics in the Field of Vision.* London: Routledge.

Rodgers, Scott. 2020a. "Journalism: An Urban Affair." In *The Routledge Companion to Urban Media and Communication,* edited by Zlatan Krajina and Deborah Stevenson, 66–75. London and New York: Routledge.

———. 2020b. "Platform Phenomenologies: Social Media as Experiential Infrastructures of Urban Public Life." In *Urban Platforms and the Future City: Transformations in Infrastructure, Governance, Knowledge and Everyday Life*, edited by John Stehlin, Michael Hodson, Julia Kasmire, Andrew Mcmeekin, and Kevin Ward, 209–222. London: Routledge.

Rodgers, Scott, and Susan Moore. 2018. "Platform Urbanism: An Introduction." *Mediapolis: A Journal of Cities and Culture* 4 (3): Available at: http://www.mediapolisjournal.com/2018/10/platform-urbanism-an-introduction/.

Roquet, Paul. 2016. *Ambient Media: Japanese Atmospheres of the Self.* Minneapolis, MN: University of Minnesota Press.

Rose, Gillian. 2010. *Doing Family Photography: The Domestic, the Public and the Politics of Sentiment.* Farnham: Ashgate.

———. 2014. "On the Relation between 'Visual Research Methods' and Contemporary Visual Culture." *The Sociological Review* 62 (1): 24–46.

———. 2016a. "Cultural Geography Going Viral." *Social & Cultural Geography* 17 (6): 763–767.

———. 2016b. "Rethinking the Geographies of Cultural 'Objects' Through Digital Technologies: Interface, Network and Friction." *Progress in Human Geography* 40 (3): 334–351.

———. 2017. “Posthuman Agency in the Digitally Mediated City: Exteriorization, Individuation, Reinvention.” *Annals of the American Association of Geographers.* 107(4): 779–793.

———. 2018. “Look InsideTM: Visualising the Smart City.” In *An Introduction to Geomedia: Spaces and Mobilities in Mediatized Worlds*, edited by Karin Fast, André Jansson, Johan Lindell, Linda Ryan Bengtsson, and Mekonnen Tesfahuney, 97–113. London: Routledge.

Rose, Gillian, and Alistair Willis. 2019. “Seeing the Smart City on Twitter: Colour and the Affective Territories of Becoming Smart.” *Environment and Planning D: Society and Space* 37 (3): 411–27.

Rose, Gillian, and Divya Tolia-Kelly, eds. 2012. *Visuality/Materiality: Images, Objects and Practices.* Farnham: Ashgate.

Rose, Gillian, Monica Degen, and Clare Melhuish. 2014. “Networks, Interfaces, and Computer-Generated Images: Learning from Digital Visualisations of Urban Redevelopment Projects.” *Environment and Planning D: Society and Space* 32 (3): 386–403.

———. 2016. “Looking at Digital Visualizations of Urban Redevelopment Projects: Dimming the Scintillating Glow Of Unwork.” In *Cities Interrupted: Visual Culture and Urban Space,* edited by Shirley Jordan and Christoph Lindner, 105–120, London: Bloomsbury.

Rose, Gillian, Parvati Raghuram, Sophie Watson, and Edward Wigley. 2020. “Platform Urbanism, Smartphone Applications and Valuing Data in a Smart City.” *Transactions of the Institute of British Geographers* 46 (1):59–72.

Rouvroy, Antoinette. 2013. “The End(s) of Critique: Data Behaviourism Versus Due Process.” In *Privacy, Due Process and the Computational Turn: The Philosophy of Law Meets the Philosophy of Technology,* edited by Mireille Hildebrandt and Katja de Vries, 143–168, Abingdon: Routledge.

Rowlands, Mark. 2010. *The New Science of the Mind: From Extended Mind to Embodied Phenomenology.* London: MIT Press.

Rubinstein, Daniel, and Katrina Sluis. 2008. “A Life More Photographic: Mapping the Networked Image.” *Photographies* 1 (1): 9–28.

Ruddick, Susan. 1996. “Constructing Difference in Public Spaces: Race, Class, and Gender as Interlocking systems.” *Urban Geography* 17 (2): 132–151.

Runge, Evelyn. 2020. “The Travels Of Photographs Within the Global Image Market: How Monopolisation, Interconnectedness, And Differentiation Shape the Economics Of Photography.” *Photographies* 13 (3): 365–384.

Ruppert, Evelyn. 2012. “Category.” In *Inventive Methods: The Happening of the Social,* edited by Celia Lury and Nina Wakeford, 36–47. Abingdon: Routledge.

Sadowski, Jathan. 2020. “Cyberspace and Cityscapes: On the Emergence of Platform Urbanism.” *Urban Geography,* January, 1–5.

Sarkar, Sreela. 2016. "Beyond the 'Digital Divide': The 'Computer Girls' of Seelampur." Feminist Media Studies 16 (6): 968–83.

Sassen, Saskia. 2013. "Does the City Have Speech?" *Public Culture* 25 (2): 209–221.

Schatzki, Theodore R. 2002. *The Site of the Social: A Philosophical Account of the Constitution of Social Life and Change.* University Park: Penn State University Press.

Schech, S. 2002. "Wired for Change: The Links between ICTs and Development Discourses." Journal of International Development 14 (1): 13–23.

Schiller, Naomi. 2020. "Urban Media as Infrastructure for Social Change." In *The Routledge Companion to Urban Media and Communication*, edited by Zlatan Krajina and Deborah Stevenson, 204-214. London and New York: Routledge.

Schmidt, Florian A. 2019. "Crowdsourced Production of AI Training Data: How Human Workers Teach Self-Driving Cars to See." Working Paper Forschungsförderung, No. 155, Hans-Böckler-Stiftung, Düsseldorf. https://d-nb.info/1197839895/34/.

———. 2020. "Unevenly Distributed." Unthinking Photography. https://unthinking.photography/articles/unevenly-distributed/.

Schwarz, Brent. 2010. "Mapping the World in 3D." *Nature Photon* 4: 429–30.

Seamon, David. 1979. *A Geography of the Lifeworld: Movement, Rest and Encounter.* London: Routledge.

Sedgwick, Eve Kosofsky. 2003. *Touching Feeling: Affect, Pedagogy, Performativity.* Duke University Press, Durham, NC.

Sennett, Richard. 2003. "Reflections on the Public Realm." In *A Companion to the City*, edited by Gary Bridge and Sophie Watson, 380–387. Oxford: Blackwell.

Shah, Nishant. 2015. "Networked Margins: Revisiting Inequality and Intersection." In *Digitally Connected: Global Perspectives on Youth and Digital Media*, edited by Sandra Cortesi and Urs Gasser, 9–11, Berkman Center Research Publication No. 2015–6, https://papers.ssrn.com/sol3/papers.cfm?abstract_id=2585686/.

Sharma, Sarah. 2014. *In the Meantime: Temporality and Cultural Politics.* Durham, NC: Duke University Press.

Shaviro, Steve. 2016. "Post-Continuity: An Introduction." In *Post-Cinema: Theorizing 21st-Century Film*, edited by Shane Denson and Julia Leyda, 52–64. Falmer: Reframe. http://reframe.sussex.ac.uk/post-cinema/.

Shaw, Joe. 2018. "Platform Real Estate: Theory and Practice of New Urban Real Estate Markets." *Urban Geography* 41 (8): 1037–1064.

Sheller, Mimi. 2015. "News Now: Interface, Ambience, Flow, and the Disruptive Spatio-Temporalities of Mobile News Media." *Journalism Studies* 16 (1): 12–26.

Shelton, Taylor. 2017. "The Urban Geographical Imagination in the Age of Big Data." *Big Data & Society* 4 (1): https://doi.org/10.1177/2053951716665129.

Shifman, Limor. 2013. *Memes in Digital Culture.* Cambridge, MA: The MIT Press.

Simondon, Gilbert. 2016. *On the Mode of Existence of Technical Objects.* Translated by Cecile Malaspina and John Rogove. Minneapolis: Univocal Publishing.

Skains, R. Lyle. 2018. "Creative Practice as Research: Discourse on Methodology." *Media Practice and Education* 19 (1): 82–97.

Srnicek, Nick. 2014. "Auxiliary Organs: An Ethics of the Extended Mind." In *The Imaginary App*, edited by Paul D Miller and Svitlana Matviyenko, 71–82. London: MIT Press.

———. 2017. *Platform Capitalism*. Cambridge: Polity.

Stacey, Noni. 2020. *Photography of Protest and Community: The Radical Collectives of the 1970s*. London: Lund Humphries.

Steinbock, Eliza. 2019. *Shimmering Images: Trans Cinema, Embodiment, and the Aesthetics of Change*. London: Duke University Press.

Stevenson, Deborah. 2003. *Cities and Urban Cultures*. Maidenhead: Open University Press.

Steyerl, Hito. 2017. *Duty Free Art: Art in the Age of Planetary Civil War*. London: Verso.

Stiegler, Bernard. 2012. "Relational Ecology and the Digital Pharmakon." *Culture Machine* 13: 1–19.

Stilgoe, Jack. 2017. "Seeing like a Tesla: How can We Anticipate Self-Driving Worlds?" *Glocalism: Journal of Culture, Politics and Innovation* 3: 1–20.

Struppek, Mirjam. 2006. "The Social Potential of Urban Screens." *Visual Communication*. 5(2): 173–188.

Subramanian, Sujatha. 2015. "From the Streets to the Web: Looking at Feminist Activism on Social Media." *Economic and Political Weekly* L: 71–78.

Sumartojo, Shanti, and Sarah Pink. 2019. *Atmospheres and The Experiential World: Theory and Methods*. London: Routledge.

Tacoli, Cecilia, and David Satterthwaite. 2013. "Gender and Urban Change." *Environment and Urbanization* 25 (1): 3–8.

Tagg, John. 1988. *The Burden of Representation: Essays on Photographies and Histories*. London.

Tarlo, Emma (2003) *Unsettling Memories: Narratives of the Emergency in Delhi*. Oakland, CA: University of California Press.

Technavio. 2019. "Stock Images Market by Application, Image Source, License Model, and Geography — Forecast and Analysis 2020-2024." https://www.technavio.com/report/stock-images-market-industry-analysis/.

Thelander, Åsa, and Cecilia Cassinger. 2017. "Brand New Images? Implications of Instagram Photography for Place Branding." *Media and Communication* 5 (4): 6–14.

Therborn, Göran. 2017. "Cities as Media of Power." *Subversive Festival*, Zagreb.

Thi Hoan, Nguyen, Arul Chib, and Ram Mahalingam. 2016. "Mobile Phones and Gender Empowerment: Enactment of 'Restricted Agency'." In *ICTD '16: Proceedings of the 8th International Conference on Information and Communication Technologies and Development* 1–10, Ann Arbor, MI.

Thrift, Nigel, and Shaun French. 2002. "The Automatic Production of Space." *Transactions of the Institute of British Geographers* 27 (3): 309–335

Thrift, Nigel. 2014. "The 'Sentient' City and What It May Portend." *Big Data & Society* 1 (1): 1–21.

Thurlow, Crispin. 2020. "Touching Taste: The Classist Intersections of Materiality, Textuality And Tactility." *Multimodal Communication* 9 (2): 1–16.

Thurlow, Crispin, Aiello, Giorgia, and Lara Portmann. 2020. "Visualizing Teens and Technology: A Social Semiotic Analysis of Stock Photography and News Media Imagery." *New Media & Society* 22 (3): 528-549.

Tierney, Thérèse F. 2013. *The Public Space of Social Media: Connected Cultures of the Network Society.* London: Routledge.

Tiidenberg, Katrin. 2018. *Selfies: Why We Love (and Hate) Them.* United Kingdom: Emerald Publishing Limited.

Tosoni, Simone, Krajina Zlatan, and Ridell Seija. 2019. "The Mediated City Between Research Fields: An Invitation to Urban Media Studies." *International Journal of Communication* 13: 5257–5267.

Toupin, Sophie. 2018. "Introduction: Radical Feminist Storytelling and Speculative Fiction: Creating New Worlds by Re-Imagining Hacking." Ada New Media (blog). 10 May 2018.

Towns, Armond R. 2020. "'The Street is the Message': Racial Violence and the White Control of Mobility" In *The Routledge Companion to Urban Media and Communication,* edited by Zlatan Krajina and Deborah Stevenson, 437–445. London and New York: Routledge.

Tuan, Yi-Fu. 1977. *Space and Place: The Perspective of Experience.* Minneapolis: University of Minnesota Press.

Uliasz, Rebecca. 2020. "Seeing like an Algorithm: Operative Images and Emergent Subjects." *AI & SOCIETY* Online First (September): 1–9.

UN Women. 2017. "Safe Cities and Safe Public Spaces: Global Results Report." New York: United Nations.

Uricchio, William. 2011. "The Algorithmic Turn: Photosynth, Augmented Reality and the Changing Implications of the Image." *Visual Studies* 26 (1): 25–35.

Vallet, Georges. 1985. "Ville et Cite: Réflexions sur les Premières Fondations Qrecques en Occident." In *L'idée de la Ville,* edited by François Guéry, 56–64. Champ Vallon: Seyssel.

Valluvan, Sivamohan. 2016. "Conviviality and Multiculture: A Post-Integration Sociology of Multi-Ethnic Integration." *Young* 24 (3): 204–221.

Van Es, Karin. 2017. "Liveness Redux: On Media and their Claim to be Live." *Media, Culture & Society* 39 (8): 1245–1256.

Velodyne Lidar. 2018a. "HDL-64E Data Sheet." HDL-64E. https://velodynelidar.com/products/hdl-64e/.

———. 2018b. “HDL-64E User’s Manual and Programming Guide.” HDL-64E. https://velodynelidar.com/products/hdl-64e/.

———. 2018c. “A Safety-First Approach to Developing and Marketing Driver Assistance Technology.” *Velodyne Lidar*, December 1, 2018. https://velodynelidar.com/wp-content/uploads/2019/08/ADAS-Whitepaperver.12.2018.pdf.

———. 2020a. “Puck: Compact, Powerful, Intelligence.” https://velodynelidar.com/products/puck/.

———. 2020b. “HDL-64E: High Definition Real-Time 3D Lidar.” https://velodynelidar.com/products/hdl-64e/.

Verhoeff, Nanna, and Clancy Wilmott. 2016. “Curating the City: Urban Interfaces and Locative Media as Experimental Platforms for Cultural Data.” In *Code and the City*, edited by Rob Kitchin and Sung-Yueh Perng, 116–129, London: Routledge.

Viqar, Sarwat. 2018. “Women’s Public Lives and Sovereign Arrangements in Karachi’s Inner City.” Gender, Place & Culture 25 (3): 416–33.

Vuolteenaho, Jani, Koen Leurs, and Johanna Sumiala. 2015. “Digital Urbanisms: Exploring the Spectacular, Ordinary and Contested Facets of the Media City.” *Observatorio* 9, http://obs.obercom.pt/index.php/obs/issue/view/48/.

Wachsmuth, David, and Alexander Weisler. 2018. “Airbnb and the Rent Gap: Gentrification through the Sharing Economy.” *Environment and Planning A: Economy and Space* 50 (6): 1147–70.

Wang, Ting-Chun, Ming-Yu Liu, Jun-Yan Zhu, Andrew Tao, Jan Kautz and Bryan Catanzaro. 2018. “High-Resolution Image Synthesis and Semantic Manipulation with Conditional GANs.” In *Proceedings of the IEEE Conference on Computer Vision and Pattern Recognition* (*CVPR*). https://ieeexplore.ieee.org/document/8579015/.

Weber, Jutta. 2010. “‘Security’ Architectures, New Ontologies and the Category of Gender. Contemporary Challenges in Feminist Technoscience Studies.” In *GenderChange in Academia: Re-Mapping the Fields of Work, Knowledge, and Politics from a Gender Perspective*, edited by Birgit Riegraf, Brigitte Aulenbacher, Edit Kirsch-Auwärter, and Ursula Müller, 217–30. Wiesbaden: VS Verlag für Sozialwissenschaften.

Weheliye, Alexander G. 2014. *Habeas Viscus: Racializing Assemblages, Biopolitics, and Black Feminist Theories of the Human*. Durham NC: Duke University Press.

Wei, Jerry. 2019. “AlexNet: The Architecture that Challenged CNNs.” Medium. https://towardsdatascience.com/alexnet-the-architecture-that-challenged-cnns-e406d5297951/.

Weizman, Eyal. 2017. *Forensic Architecture: Violence at the Threshold of Detectability*. New York: Zone Books.

Weltevrede, Esther, Anne Helmond, and Carolin Gerlitz. 2014. “The Politics of Real-Time: A Device Perspective on Social Media Platforms and Search Engines.” *Theory, Culture & Society* 31 (6): 125–50.

Wentworth, Jane. 2018. "Culture Mile: The Creative Heart of the Square Mile." Presentation at Jane Wentworth Associates, Nov 11, 2018.

Werner, Alex. 2017. "Foreword: The Changing Feel of Smithfield." In *The Changing Feel of Smithfield: Exploring Sensory Identities and Temporal Flows*, by Monica Degen, Camilla Lewis, Astrid Swenson, and Isobel Ward. London: Brunel University.

Whitzman, Carolyn, Caroline Andrew, and Kalpana Viswanath. 2014. "Partnerships for Women's Safety in the City: 'Four Legs for a Good Table'." *Environment and Urbanization* 26 (2): 443–56.

Wilken, Rowan. 2019. *Cultural Economies of Locative Media*. Oxford: Oxford University Press.

Willems, Wendy. 2019. "'The Politics of Things': Digital Media, Urban Space, and the Materiality of Publics." *Media, Culture and Society* 41 (8): 1–18.

Williams, Raymond. 1990. *Television: Technology and Cultural Form*. London: Fontana.

Willis, Katherine, and Alessandro Aurigi. 2018. *Digital and Smart Cities*. Abingdon: Routledge.

Wilmott, Clancy. 2016. "In-between Mobile Maps and Media: Movement." *Television & New Media* 18 (4): 320–35.

Wilson, Matthew W. 2011. "'Training the Eye': Formation of the Geocoding Subject." *Social and Cultural Geography* 12 (4): 357–76.

———. 2014. "Continuous Connectivity, Handheld Computers, and Mobile Spatial Knowledge." *Environment and Planning D: Society and Space* 32 (3): 535–55.

Winter, Rachel, and Anna Lavis. 2019. "Looking, But Not Listening? Theorizing the Practice and Ethics of Online Ethnography." Journal of Empirical Research on Human Research Ethics 15 (1–2): 55–62.

Wirth, Louis. 1938. "Urbanism as a Way of Life." *American Journal of Sociology* 44 (1): 1–24.

Withers, Deborah. 2015. *Feminism, Digital Culture and the Politics of Transmission*. London: Rowman & Littlefield International.

Zuboff, Shoshana. 2019. *The Age of Surveillance Capitalism: The Fight for A Human Future at the New Frontier of Power*. London: Profile.

Zukin, Sharon. 1991. *Landscapes of Power: From Detroit to Disney World*. London: University of California Press.

———. 1995. *The Cultures of Cities*. Malden: Blackwell.

Index

References to figures are indicated in *italics*. References to footnotes consist of the page number followed by the letter 'n' followed by the number of the note, e.g. 215n3.

For Product Safety Concerns and Information please contact our EU representative GPSR@taylorandfrancis.com Taylor & Francis Verlag GmbH, Kaufingerstraße 24, 80331 München, Germany